IoT and Advanced Intelligence Computation for Smart Agriculture

Smart agriculture is an approach to maintaining nature without compromising the basic needs of future generations while at the same time improving the efficiency of farming. The main achievements of smart agriculture in terms of sustainable agriculture are crop rotation, controlling nutrient deficiencies in crops, pest, and disease control, recycling, and water harvesting, leading to a safer environment overall. Living organisms depend on the nature of biodiversity and are exposed to pollution due to waste emissions, use of fertilizers and pesticides, degraded dead plants, and so on. The emission of greenhouse gases affects plants, animals, humans, and the environment; hence, this necessitates a better environment for living organisms

The purpose of this book is to provide a comprehensive overview of the latest advancements, challenges, and potential applications of artificial intelligence (AI) technology and the Internet of Things (IoT) in the future of intelligent agriculture. The book is primarily focused on equipping younger researchers, graduates, and professionals in the industry with the necessary knowledge to understand the advantages of AI technology, machine learning, and data analytics methods in improving current agricultural practices.

Key features include the following:

- The book showcases the latest advancements in AI and smart agriculture technologies
- The text emphasizes sustainable practices supported by AI, highlighting how technology can enhance productivity while minimizing environmental impact
- Readers will learn how to harness big data and analytics to drive informed decision-making and optimize their agricultural yields

IoT and Advanced Intelligence Computation for Smart Agriculture

Edited by Mourade Azrour,
Jamal Mabrouki, and Sultan Ahmad

CRC Press
Taylor & Francis Group
Boca Raton London New York

CRC Press is an imprint of the
Taylor & Francis Group, an **informa** business

First edition published 2026
by CRC Press
2385 Executive Center Drive, Suite 320, Boca Raton, FL 33431

and by CRC Press
4 Park Square, Milton Park, Abingdon, Oxon, OX14 4RN

CRC Press is an imprint of Taylor & Francis Group, LLC

Library of Congress Cataloging-in-Publication Data
Names: Azrour, Mourade editor l Mabrouki, Jamal editor l Ahmad, Sultan (Computer scientist) editor
Title: IoT and advanced intelligence computation for smart agriculture / edited by Mourade Azrour, Jamal Mabrouki, and Sultan Ahmad.
Description: First edition. l Boca Raton, FL : Taylor and Francis, 2026. l Includes bibliographical references and index.
Identifiers: LCCN 2025005966 (print) l LCCN 2025005967 (ebook) l ISBN 9781032852386 hardback l ISBN 9781032864662 paperback l ISBN 9781003527664 ebook
Subjects: LCSH: Agricultural informatics l Agricultural innovations l Artificial intelligence—Agricultural applications l Internet of Things
Classification: LCC S494.5.D3 I59 2026 (print) l LCC S494.5.D3 (ebook) l DDC 338. 10285—dc23/eng/20250403
LC record available at https://lccn.loc.gov/2025005966
LC ebook record available at https://lccn.loc.gov/2025005967

ISBN: 978-1-032-85238-6 (hbk)
ISBN: 978-1-032-86466-2 (pbk)
ISBN: 978-1-003-52766-4 (ebk)

DOI: 10.1201/9781003527664

Typeset in Times
by Apex CoVantage, LLC

Contents

Editors

Mourade Azrour received his PhD from the Faculty of Sciences and Techniques, Moulay Ismail University of Meknès, Morocco. He received an MS in computer and distributed systems from the Faculty of Sciences, Ibn Zouhr University, Agadir, Morocco, in 2014. Dr. Azrour currently works as Professor of Computer Science in the Department of Computer Science, Faculty of Sciences and Techniques, Moulay Ismail University of Meknès. His research interests include authentication protocol, computer security, Internet of Things, smart systems, machine learning, and so on. Dr. Azrour is a member of scientific committees of numerous international conferences. He is also a reviewer of various scientific journals. He has published more than 127 scientific papers and book chapters. He has edited many scientific books: *IoT, Machine Learning and Data Analytics for Smart Healthcare*; *Blockchain and Machine Learning for IoT Security*; *IoT and Smart Devices for Sustainable Environment*; and *Advanced Technology for Smart Environment and Energy*. He has also served as guest editor of journals *EAI Endorsed Transactions on Internet of Things, Tsinghua Science and Technology, Applied Sciences MDPI*, and *Sustainability MDPI*.

Jamal Mabrouki is a researcher and expert in water science and technology. He is also an engineer in environment and climate. Mabrouki is working on a project on migration and water and the role of water governance in migration policy in Africa in cooperation with MedYWat and the World Bank. He is currently a researcher for the Environment and Climate Program at ECOMED in Morocco, where he started as the Coordinator of the project "Adaptation of Citizens to Climate Change."

Sultan Ahmad received his PhD from Glocal University and Master of Computer Science and Applications degree from the prestigious Aligarh Muslim University, India, with distinction. Presently, he is working as faculty in the Department of Computer Science, College of Computer Engineering and Sciences, Prince Sattam Bin Abdulaziz University, Alkharj, Saudi Arabia. He is also an Adjunct Professor at Chandigarh University, Gharuan, Punjab, India. He has more than 15 years of teaching and research experience. He has around 80 accepted and published research papers and book chapters in reputed Science Citation Index Expanded (SCIE), Emerging Sources Citation Index (ESCI), and SCOPUS-indexed journals and conferences. He has an Australian Patent in his name. He has authored four books, which are available on Amazon. His research areas include distributed computing, big data, machine learning, and the Internet of Things. He has presented his research papers at many national and international conferences. He is a member of Institute of Electrical and Electronics Engineers (IEEE), International Association of Computer Science and Information Technology (IACSIT), and the Computer Society of India.

Contributors

Hikmat A.M. Abdeljaber
Faculty of Information Technology
Applied Science Private University
Amman, Jordan

Mazhar Afzal
Glocal University
Saharanpur, Uttar Pradesh, India

Badraddine Aghoutane
Moulay Ismail University of Meknès
Meknès, Morocco

Said Agoujil
Moulay Ismail University of Meknès
Meknès, Morocco

Sultan Ahmad
Prince Sattam Bin Abdulaziz University
Al-Kharj, Saudi Arabia
and
Chandigarh University
Mohali, Punjab, India

Farzana Akter
Rajshahi University of Engineering &
    Technology
Rajshahi, Bangladesh

Aleem Ali
Chandigarh University
Mohali, Punjab, India

Mourade Azrour
Moulay Ismail University of Meknès
Meknès, Morocco

M.J. Carmel Mary Belinda
Saveetha School of Engineering
Saveetha Institute of Medical and
    Technical Sciences
Chennai, Tamil Nadu, India

Said Benkirane
Cadi Ayyad University
Marrakesh, Morocco

Manal Benzyane
Moulay Ismail University of Meknès
Meknès, Morocco

Shivani Bhardwaj
Shoolini University
Solan, Himachal Pradesh, India

Mohamed Khalifa Boutahir
National School of Artificial
    Intelligence and Digitalization
Berkane, Morocco

Sreedeep Dey
University of Calcutta
Kolkata, West Bengal, India

Ahmad El Allaoui
Moulay Ismail University
    of Meknès
Meknès, Morocco

Omaima El Bahi
Moulay Ismail University of Meknès
Meknès, Morocco

Ahmed El Youssefi
Moulay Ismail University of Meknès
Meknès, Morocco

Yousef Farhaoui
Moulay Ismail University of Meknès
Meknès, Morocco

Rohit Ghatuary
Lovely Professional University
Phagwara, Punjab, India

Azidine Guezzaz
Cadi Ayyad University
Marrakesh, Morocco

Gauarv Gupta
Shoolini University
Solan, Himachal Pradesh, India

A.K.M. Bahalul Haque
Software Engineering, LENS, LUT
   University
Lappeenranta, Finland

Md. Alimul Haque
Department of Computer Science
Veer Kunwar Singh University
Arrah, Bihar, India

Abdelaaziz Hessane
Moulay Ismail University of Meknès
Meknès, Morocco

Al Amin Islam Ridoy
Rajshahi University of Engineering &
   Technology
Rajshahi, Bangladesh

Sapna Jarial
Lovely Professional University
Phagwara, Punjab, India

Anas Kabbori
Cadi Ayyad University
Marrakesh, Morocco

Rupesh Kaushik
Lovely Professional University
Phagwara, Punjab, India

Hajar Khabouche
University Mohammed V in Rabat
Rabat, Morocco

Amit Kotiyal
Lovely Professional University
Phagwara, Punjab, India

K.A. Varun Kumar
SRM Institute of Science and Technology
Kattankulathur, Tamil Nadu, India

Rajneesh Kumar
Sher-e-Kashmir University of Agricultural
   Sciences and Technology (SKUAST–K)
Jammu, Jammu and Kashmir, India

Jamal Mabrouki
University Mohammed V in Rabat
Rabat, Morocco

Ashish Kumar Mourya
ABES Institute of Technology
Ghaziabad, Uttar Pradesh, India

Ali Omari Alaoui
Moulay Ismail University
   of Meknès
Meknès, Morocco

Shagufta Praveen
JB Institute of Technology
Dehradun, Uttarakhand, India

Youssef Qaraai
Moulay Ismail University
   of Meknès
Meknès, Morocco

Seema Rani
Amity University
Greater Noida, Uttar Pradesh, India

Subhasis Roy
University of Calcutta
Kolkata, West Bengal, India

Miloudia Slaoui
University Mohammed V in Rabat
Rabat, Morocco

Deepa Sonal
Patna Women's College
Patna, Bihar, India

Mukesh Tiwari
Shoolini University
Solan, Himachal Pradesh, India

Abhishek Tomar
Shoolini University
Solan, Himachal Pradesh, India

Shafqat Ul Ahsaan
NIMS University
Jaipur, Rajasthan, India

T. Anstey Vathani
Vel Tech Rangarajan Dr. Sagunthala R&D
 Institute of Science and Technology
Chennai, Tamil Nadu, India

K.A. Vinodhini
Vel Tech Rangarajan Dr. Sagunthala
 R&D Institute of Science and
 Technology
Chennai, Tamil Nadu, India

Aasim Zafar
Aligarh Muslim University
Aligarh, Uttar Pradesh, India

1 Smart Farming
An IoT-Enhanced and Interpretable Optimal Crop Selection Technique Based on Soil and Environmental Data

Al Amin Islam Ridoy, A.K.M. Bahalul Haque, Farzana Akter, and Mourade Azrour

1.1 INTRODUCTION

Agriculture is widely recognized as the world's most ancient and fundamental activity. It supplies the necessary food and livestock to feed the billions of people living on Earth [1]. Agriculture supports diverse economies by providing nutrition, employment, and essential substances for various industries [2]. Farmers have relied on regional or local information and traditional wisdom of past centuries to decide what and when to grow crops, what amount to sow, and how to allocate available assets throughout the generations [3]. In recent decades, there has been a significant and rapid change in the planet's climate due to global warming [4]. As a result, the environmental conditions (temperature, rainfall, humidity) surrounding the farming field are different from before. Moreover, the soil characteristics also change after each successful crop cultivation. However, farmers do not consider these issues when they decide which crop they want to grow in a particular season, and often they don't know how to select an optimal crop based on the soil and environmental conditions. So, traditional approaches struggle to address the unique requirements of different crops [5]. If a crop has been selected for cultivation without adequate thinking, it could lead to harvest failure and endanger the financial stability of the farmer's family. This could be one of the causes that contribute to the quitting of farming, migration to urban areas, and a rise in farmer suicide cases, which have been confirmed in the mainstream media [6].

On the other hand, around 10 billion people will call this planet home by the year 2050, significantly increasing the need for food. That's why there is an urgent requirement to boost present agricultural food production by over 70% before the year 2050 to sustain the expanding global population adequately. Yet, according to

DOI: 10.1201/9781003527664-1

1

the experts, there has been a decline in land fertility over time, which has impacted crop production [7]. In addition, as urbanization keeps growing, there will undoubtedly be a significant reduction in agricultural land. Therefore, the requirement for manual labor, diminishing arable land, soil degradation, and rising capital expenses are making it even more challenging to fulfill the rising demand for agricultural products. These issues raise concerns about the feasibility of current farming practices [8]. A modern agricultural strategy called Smart Farming integrates conventional farming practices with AI, IoT, wireless communication, and robotics to enhance the profitability and productivity of traditional farming practices [9–12]. Automated data-gathering systems incorporating ML algorithms have made data-driven decisions possible for farmers. These decisions encompass various aspects such as what type of crops to cultivate, when and how much to water, what pesticides to apply, and predicting future yields. The smart farming approach opened many opportunities for researchers and developers to discover innovative novel strategies that will produce huge amounts of high-quality crops with fewer resources, and selecting the appropriate crop for cultivation is the first step toward it.

Numerous studies have been carried out on this topic, all based on machine learning (ML) or deep learning (DL) approaches. Farmers and agronomists are the intended individuals as they are directly related to the harvest production procedure. However, they often need help accepting or trusting AI models' suggestions, especially when those suggestions involve non-interpretable black-box models [13]. Here, XAI can play a crucial role as it reveals the reasons behind a prediction in an easy-readable format. However, most of the research work [14–16] didn't mention Explainable Artificial Intelligence (XAI) technology, which hinders the widespread adoption of AI. Furthermore, recommendation accuracy is another concerning term, as one bad recommendation can cause heavy damage to farm and a country's economy. Yet, the majority of academic work [17–20] in crop recommendations holds relatively low accuracy, which is quite alarming. Furthermore, deploying IoT gadgets is regarded as one of the top four challenges in the future of agricultural innovation as these devices offer instantaneous data gathering, data monitoring, and resource management to reduce production costs and help increase productivity. However, several studies don't discuss the implementation of IoT devices for real-time data gathering and monitoring systems [21–24].

This study introduces a novel paradigm that recommends the most appropriate crops for farmers by inspecting three environmental parameters (temperature, humidity, and rainfall) and four soil parameters (N, P, K, pH). SHAP and LIME, two prominent XAI tools that provide a pictorial representation of feature contribution, have been used to explain the decision of the AI model. Furthermore, this research framework allows continuous data collection and monitoring by leveraging IoT devices with a user-friendly communication interface to obtain personalized recommendations. The key contributions of this chapter are as follows:

- Offers a customized and easily accessible data-driven crop recommendation system to suggest to farmers only the most appropriate crops to cultivate in a particular land.

- Presents a comparison between several machine learning algorithms and deep learning techniques to determine the most accurate (99.55%) AI model with outstanding performance metrics.
- Cutting-edge IoT technology is utilized for automatic data retrieval of multiple soil and environmental characteristics from the targeted field where farmers intend to cultivate their crops.
- Integration of XAI tools to provide significant insights and enhance the interpretability, transparency, and widespread applicability of AI in the agriculture sector.
- Comparison between relevant state-of-the-art works to validate and distinguish this work from others.

The overall architectural workflow of this research is depicted in Figure 1.1. The subsequent sections of this chapter have been organized systematically. First,

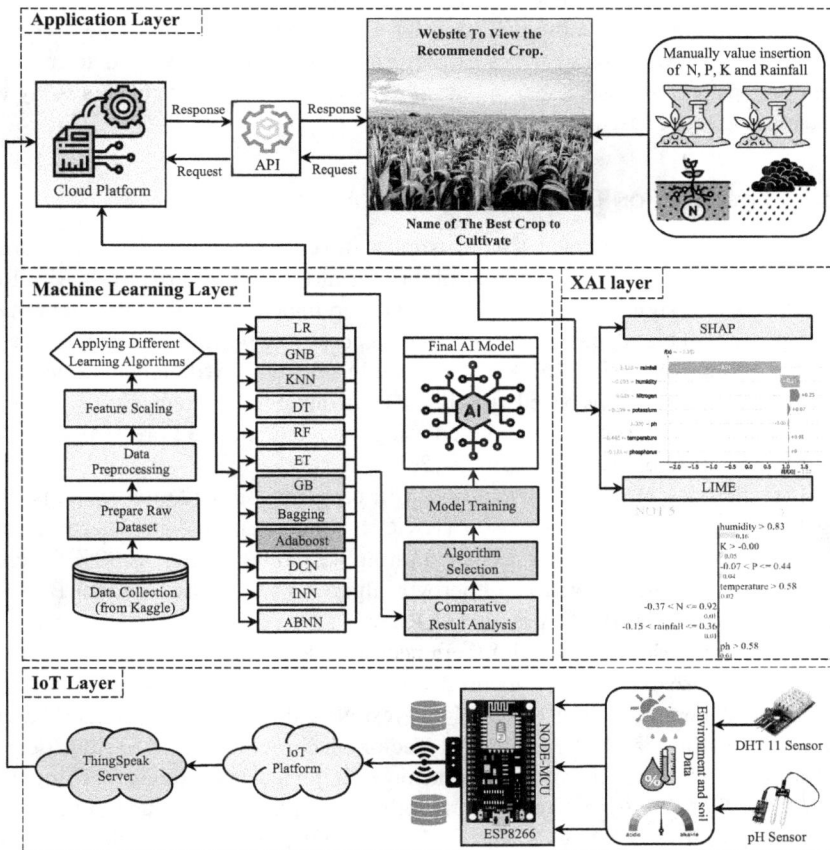

FIGURE 1.1 Architectural layout of the intended framework.

this chapter portrays a review and analysis of several of the most recent well-published literature on crop recommendations. The next objective is to pinpoint areas requiring additional investigation and propose research subjects by examining gaps and raising research questions. A detailed overview of the key elements of designing this framework has been discussed. Thereafter, the corresponding methodology has been employed to build the AI model. Later, this chapter proceeded to present comparative results between different ML algorithms and DL techniques in terms of performance and computational efficiency. Subsequently, some graphical depictions of XAI and how IoT gadgets have been employed for collecting real-time data have been presented. Finally, the contributions to this study compared to existing literature have been provided. In the end, an overview of this study and encouraging possibilities for future research are discussed.

1.2   RELATED WORKS

This section has been divided into two sections. In Section 1.2.1, several of the most recent literature on crop selection have been reviewed to provide a conceptual foundation. An investigation was performed on the literature to determine the research gaps and to raise research questions. The following parts of this study will be answering those research questions.

1.2.1   LITERATURE ON CROP RECOMMENDATION SYSTEMS

Recent progress has integrated XAI principles into crop recommendation systems to enhance transparency and comprehensibility for farmers. Similarly, the XAI-CROP algorithm was designed to utilize XAI to gain an understanding of the recommendation process. This algorithm achieved an accuracy of 94.15% and high scores in other performance metrics [25]. In another study, an ANN (artificial neural network) was applied to select crops based on soil and environmental characteristics. Also, the yield of the selected crops was predicted using the RF (Random Forest) algorithm to ensure the maximum profit for farmers [26].

ML techniques have been widely used in crop recommendation systems for a long time. Kernel Ridge Regression (KRR) has been utilized to determine how well it performs against several crops, attaining R2 values as high as 99% for crops such as potatoes and wheat [27]. Other ML algorithms, such as Gradient Boosting Machine (GBM), Logistic Regression (LR), and Decision Trees (DT), have proved themselves quite useful by providing accuracy score ranging from 65% to 85% [28]. Several research articles focused on performance comparisons of different algorithms. A comparison of KNN (K-Nearest Neighbors), DT, RF, XGB (Extreme Gradient Boosting), and SVM (Support Vector Machine) indicated various degrees of accuracy. KNN and RF consistently achieved high accuracy rates, reaching up to 96.36% and 97.18%, respectively [29]. Another research study included five distinct algorithms, such as KNN, DT, RF, GNB, and XGB, and recorded their accuracy rates. Out of all the options, XGB demonstrated outstanding performance [30].

Meanwhile, deep learning techniques based on neural network models are also employed. These technologies are designed to provide precise and practical recommendations, helping farmers make more informed decisions when picking suitable crops [31, 32]. Furthermore, academics have investigated hybrid and ensemble ML approaches to make the predictions more accurate. Techniques like IDCSO (Improved Distribution-Based Chicken Swarm Optimization) with "WLSTM" (Weight-based Long Short-Term Memory) [33] and "MSVM-DAG-FFO" (Multi-Class Support Vector Machine with Directed Acyclic Graph-based Feature Fusion Optimization) [34] have demonstrated extraordinary performance. All this literature suggests a growing inclination to combine sophisticated ML algorithms with XAI to create crop recommendation systems that are interpretable, highly accurate, and have an easy-to-use interface to serve every category of farmers. Table 1.1 offers

TABLE 1.1

An Overview of the Various Approaches for Crop Recommendation Systems

| | | | | Presence of | |
| | | | | User | IoT |
Year	Author	Contribution and Discussion	XAI	Interface	Devices
2024	Shams et al. [25]	Developed an algorithm named XAI-CROP that harnesses the XAI principle and gives farmers a clear understanding of the suggestion procedure, exhibiting an accuracy of 94.15%, MSE score of 0.9412, and MAE score of 0.9874	✓	✗	✗
2024	Bhola et al. [26]	The study consists of crop selection and yield prediction. The authors applied ANN and achieved 99.10% accuracy for crop selection. On the other hand, RF was used for yield prediction that exhibited $0.99R^2$ and 9.7e1 RMSE	✗	✗	✗
2023	Hasan et al. [27]	The proposed system is based on the ensemble ML technique KRR to analyze and evaluate the outcomes for a set of five different crops. The MSE for Aus is 0.009 with an R2 value of 99%. Aman has an MSE of 0.92 with an R2 of 90%. Boro has an MSE of 0.246 and an R2 of 99%. Wheat has an MSE of 0.062 and an R2 of 99%. Lastly, Potato has an MSE of 0.016 and an R2 of 99%	✗	✗	✗
2023	Kumar et al. [28]	Gradient Boosting Machine (GBM), LR, and DT were used with an accuracy score of 85%, 75%, and 65%, respectively	✗	✗	✗
2023	Dolli et al. [30]	Five distinct algorithms were employed, and their corresponding accuracy scores are as follows: KNN 94.6%, DT 88.18%, RF 93.23%, GNB 90%, and XGB with a score of 96.72%	✗	✗	✗

(Continued)

TABLE 1.1   (Continued)
An Overview of the Various Approaches for Crop Recommendation Systems

				Presence of	
Year	Author	Contribution and Discussion	XAI	User Interface	IoT Devices
2023	Musanase et al. [31]	Developed a neural network model to recommend crops from a pool of nine distinct categories, with an accuracy rate of 97%	✗	✗	✓
2023	Kiruthika and Karthika [33]	Introduces a technique based on IDCSO-WLSTM to predict and recommend crops with a high level of accuracy (92.68%), precision (90.88%), and recall (91.98%)	✗	✗	✓
2023	Senapaty et al. [34]	The proposed algorithm was MSVM-DAG-FFO and it achieved an accuracy of 97.3%. The study was performed only on four different crops (cotton, groundnut, maize, and rice)	✗	✗	✓
2022	Thilakarathne et al. [29]	Five different algorithms were used to recommend crops, including KNN, DT, RF, XGB, and SVM, with an accuracy of 96.36%, 86.64%, 97.18%, 95.62%, and 87.38%, respectively	✗	✓	✓
2022	N. and Choudhary [32]	Employed a deep learning model in conjunction with SVM, GNB, KNN, DTT, and LDA. The deep learning model obtained the most precise results of 87% when applied to the nutrition dataset	✗	✗	✗

a comprehensive analysis of those kinds of literature, including the methods used and their efficiency, along with whether other modern technologies were present.

1.2.2 RESEARCH GAPS AND QUESTIONS

From the literature review, many significant research gaps have been found. Initially, not enough studies specifically examined the reliability or interpretability of the black box model. The lack of explainability in ML hinders the widespread implementation of artificial intelligence. The potential for artificial intelligence to make incorrect decisions may outweigh its benefits of being precise, quick, and effective in decision-making if it cannot provide a clear explanation related to the context of precision agriculture. As a result, the application and scope of AI would be significantly diminished. Explainable AI would be highly beneficial in these cases and enhance the acceptance rate of AI-based predictions in farming. Beyond that, most research did not demonstrate significant interest in reducing the allocation of resources to a specific crop. Efficient utilization of resources can enhance the economic viability of farming. Furthermore, most research did not prioritize implementing ML models in a user-friendly interface that may directly help farmers of every kind, from small-scale to large-scale. So, the absence of an intuitive interface for end users undermines the objective of smart farming. Most importantly, the greater the model's accuracy, the more advantageous it is for users, leading to enhanced results and effectiveness. However, in most research, there was a lack of a more precise and advanced model incorporating IoT technologies. The integration of explainable AI, IoT technology, and a user-friendly interface will significantly boost the domain of smart farming. According to the existing gaps in research, the following questions are suggested for further investigation:

- How can a highly accurate AI model be used to recommend crops?
- ML or DL, which approach would provide a precise and computationally efficient AI model?
- What strategies can be employed to collect and monitor instantaneous data from the farming field?
- Why should the agricultural community trust the opaque and complex AI model's recommendations, and how can a transparent or interpretable system be introduced?
- How can agronomists and small- or large-scale farmers take direct advantage of the smart farming approach backed by AI?

1.3 METHODOLOGICAL FOUNDATION

In this section, the fundamental concepts of this framework, along with the algorithms for designing the whole crop recommendation system and the working algorithm of SHAP and LIME, will be discussed.

1.3.1 SMART FARMING

Precision agriculture, sometimes called smart farming, is a contemporary farming strategy that employs modern equipment to systematically observe, measure, and resolve issues in a sophisticated manner. The goal is to improve the effectiveness of on-site management and data-driven decisions concerning crop production. Furthermore, the utilization of GPS and GNSS for accurate field mapping and navigation is quite normal [35]. Smart farming also employs IoT sensors for instantaneous data collection and monitors soil moisture, temperature, nutrients, and other environmental factors. Moreover, this agricultural procedure conducts data analysis and employs the gathered data to make well-informed decisions such as recommending suitable crops, predicting yields, and suggesting the amount of fertilizer while also quantifying the quantity [36]. This utilizes advanced technologies like automation and robotics, AI and ML, big data, and cloud computing. Its purpose is to enhance the sustainability and improvement of farming practices [37].

IoT, GPS, and sensors provide real-time data on soil, crops, and the environment, as well as geographic information system (GIS) data for mapping and monitoring [38]. ML and DL approaches are required when processing and analyzing data, providing insights into the decision support system [39]. Efficiency in communication and distribution may be achieved through mobile applications and web platforms that offer rapid data retrieval and feedback. This study also contributes to the intelligent farming approach as it contains IoT, AI, and an interactive user interface that recommends crops to cultivate. This entire process establishes an integrated system for managing the environment and agricultural resources. The total framework development of this study has been provided in Algorithm 1.1.

Algorithm 1.1: XAI Enhanced Crop Recommendation System Development

Input: Dataset with features and target class.

Output: A highly accurate AI model.

Output: XAI plots of SHAP and LIME for explainability and transparency.

---

a. **Data loading and preprocessing**:
   - Load the entire dataset on the pandas DataFrame. Check for the null value and if the null value arises then handle them logically.
   - Create a dictionary to map the crop names to numerical format.
b. **Feature engineering and data split**:
   - Scale the entire dataset and separate the 7 features from the target class.
   - Split the dataset into training and testing sets, maintaining the percentage of 80% and 20%, respectively.
c. **Apply machine learning algorithm**:
   - Use Logistic Regression (LR), Gaussian Naive Bayes (GNB), K-Nearest Neighbor (KNN), DT (Decision Tree), ET (Extra Tree), Random Forest (RF), Bagging and XGB.
   - Generate the confusion matrices and other performance metrics for all the algorithms.
d. **Apply deep learning techniques**:
   - Use deep and cross-neural networks (DCN), Attention-based neural networks (ABNN), and improved neural networks (INN), some advanced deep learning techniques, and fine-tune them.
   - Generate the confusion matrices and other performance metrics for all the algorithms.
e. **Model selection**:
   - Select one ML algorithm or DL technique based on the overall performance comparison, on which the final AI model will be trained.
   - Integrate different XAI tools into the finally selected AI model.

f. SHAP interpretation and visualization:
- Use SHAP explainer to generate SHAP values and explanation for a particular instance.
- Generate different types of plots to analyze the explanations better.

g. LIME interpretation and visualization:
- Perturb the instance and generate an explanation mimicking the actual model.
- Present the explanations in human-readable format.

h. Implementation of IoT and website:
- Use several IoT sensors to collect instantaneous data from the targeted field.
- Transfer and store the data into a server for monitoring.
- Place the final AI model at the website's backend (user interface), and IoT-gathered data will be automatically retrieved from the server and placed on the website.
- Fill up the other empty databoxes and click the recommendation button.
- The most appropriate crop will be visible along with the explanation.

1.3.2 EXPLAINABLE AI

Significant progress in artificial intelligence (AI) during the past decade has allowed for developing and deploying algorithms to handle various issues. Consequently, there has been an increase in system complexity and the prevalence of opaque, non-transparent AI models [40]. As reliance on intelligent machines increases, there is a corresponding need for more transparent and interpretable models. Furthermore, the capacity to comprehensively explain the model has become the benchmark for establishing confidence and implementing artificial intelligence systems in critical fields such as healthcare and agriculture. To create an XAI-powered service, it is crucial to have a solid grasp of the core ideas in XAI and the relevant programming frameworks [41]. In this study, SHAP and LIME have been used as XAI tools. The work processes SHAP and LIME in this study have been described in the next sections.

1.3.2.1 SHAP

Shapley values, a metric of feature importance, are used in the SHAP method for evaluating the relevance of features in a single decision made by the complex AI model [42–44]. It offers a standard procedure for determining the significance of each feature. Cooperative game theory, which attempts to equitably distribute a group reward among participants according to their performance, is the primary concept of Shapley values. Within the SHAP framework, the "players" refer to the dataset's features, while the "payoff" corresponds to the prediction generated by the model. The payoff should be distributed fairly according to each player's contribution in a particular match (model's prediction). SHAP guarantees that the contribution of each feature is reviewed independently. SHAP establishes the model's mean estimate (base value) and determines each feature's proportional impact in magnitude and direction on shifting the model's final predicted score from the base value. It can provide explanations at both the local and global levels. The operational algorithm of how SHAP is applied in this research to analyze and interpret the final ML model's decision is given in Algorithm 1.2.

Algorithm 1.2: SHAP Explanation Generation

Input: A complete dataset on which the model is trained.

Input: Trained classification model $f(x)$.

Input: A specific instance x_i.

Output: Shapley values ϕ_j presenting the contribution of each feature.

Output: Visualization of SHAP values using different plots.

1.3.2.2   LIME

LIME's objective is to offer understandable explanations for individual predictions generated by AI models that are inherently challenging to comprehend due to their complex nature [45]. Instead of explaining the model's global behavior, LIME prioritizes explaining the prediction process of a model for a specific data point. An individual data sample is altered many times by modifying specific feature values, and the resulting impact on the output is observed [46]. The operational algorithm of LIME in this research is described in Algorithm 1.3.

Algorithm 1.3: LIME Explanation Generation

Input: Dataset that has been used to train the model, X_{train}

Input: Instance of interest, x_i

Input: Black-box model whose predictions are to be explained, f

Output: Interpretation of the model's decision in a human-readable format.

1.4   PROPOSED AI MODEL TRAINING

The AI model used in this study underwent several stages, starting with dataset collection and ending with model implementation. This section will review the

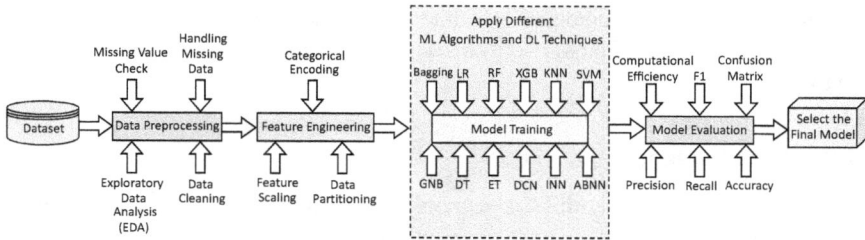

FIGURE 1.2   Workflow for training, evaluation, and selection of the AI model.

TABLE 1.2
Brief Overview of the Dataset Utilized in the Study

Feature's Name	Description	Target Class or Label
N	The ratio of nitrogen content in soil	"Rice," "Maize," "Jute," "Cotton,"
P	The ratio of phosphorus content in soil	"Coconut," "Papaya," "Orange,"
K	The ratio of potassium content in soil	"Apple," "Muskmelon," "Watermelon," "Grapes," "Mango," "Banana,"
Temperature	Weather temperature in Celsius	"Pomegranate," "Lentil," "Black gram,"
Humidity	Relative humidity in percentage	"Mungbam," "Moth bean," "Pigeon
pH	pH value of the soil	peas," "Kidney beans," "Chickpea,"
Rainfall	Rainfall in mm	"Coffee"

step-by-step process of creating the AI model from scratch. The complete procedure for constructing the ML model employed by this research is illustrated in Figure 1.2.

1.4.1 DATASET

The initial stage of developing this crop recommendation tool includes the collection of appropriate datasets. The dataset that is utilized to train a highly accurate ML model has been taken from Kaggle [47]. This dataset has been selected for model training since it contains essential parameters of the environment and soil, and the dataset is appropriately balanced. It comprises 2,200 records of 22 crops and 7 features, including environmental and soil parameters. Table 1.2 provides a brief overview of the dataset.

The soil parameters N, P, K, and pH are necessary for determining the soil characteristics. Each crop requires the appropriate concentration of these four essential elements, which are vital to the progress and growth of plants. Nitrogen affects the development of leaves, while phosphorus boosts the growth of roots, flowers, and fruits. The plant can absorb water more efficiently and resist dangerous pests and frosts with the help of potassium. Plants' access to soil nutrients and the reactions of those nutrients to one another both are affected by soil pH. Moreover, environmental parameters play a significant role in crop selection, as not every crop is meant for every environment. The ecological parameters are temperature, humidity, and rainfall. For any given plot of land, these characteristics can help determine the optimal crop to grow.

1.4.2 DATA PREPROCESSING

The data preprocessing module is responsible for analyzing and processing the acquired data. It consists of feature description, null value check, handling the null values or missing values, and exploratory data analysis such as feature distribution and correlation of features. First of all, the presence of the null values in the dataset and if any rows had any unwanted data were checked with the help of the Pandas library in Python. Fortunately, there was nothing like this. N, P, and K have an integer data type, whereas temperature, humidity, pH, and rainfall consist of floating-point data. The target, the name of 22 crops, was defined as an object here. After checking the missing values, the dataset analysis procedure starts. Figure 1.3 illustrates a series of density plots for various features in the crop recommendation dataset. Every plot demonstrates the dispersion of values for a certain characteristic. The distribution of nitrogen has a bimodal pattern, characterized by two distinct peaks centered at 25 and 85. This suggests that there are two anticipated nitrogen levels in the soil samples. The phosphorus concentration also has a bimodal distribution, with prominent peaks occurring at approximately 30 and 55. The distribution of phosphorus exhibits less uniformity than nitrogen, suggesting variability in phosphorus levels.

Similar to nitrogen and phosphorus, the distribution of potassium content is also bimodal, with two peaks observed around 25 and 200. The initial peak exhibits more elevation and distinctness than the subsequent peak, indicating that most samples possess potassium levels near 50. The temperature distribution exhibits a predominantly single-peaked pattern, with the highest concentration of samples occurring within the 25–30°C range. The humidity distribution is bimodal, with prominent peaks at around 65% and 95%. These findings suggest that the dataset contains two prevalent humidity

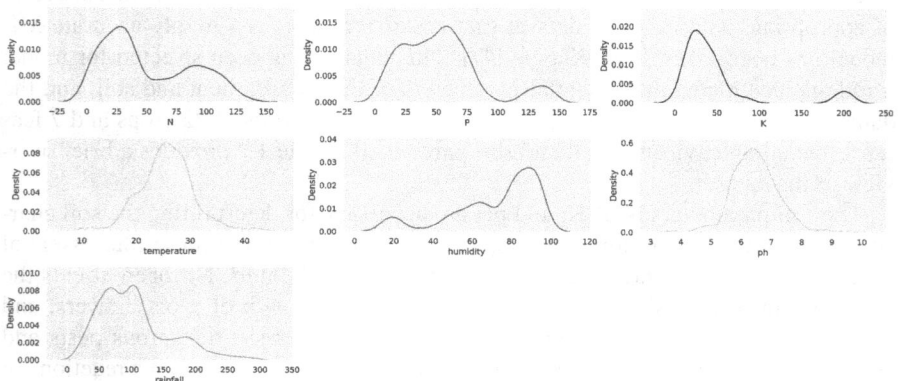

FIGURE 1.3   Complete breakdown of feature distribution throughout the dataset.

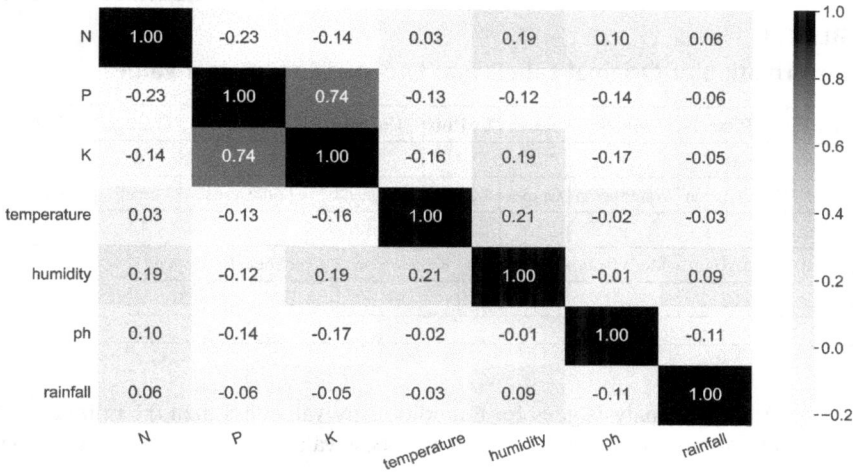

FIGURE 1.4   Insights of the correlations between features in the dataset.

levels. The pH distribution exhibits a symmetrical bell-shaped curve, with its peak centered at pH 6. Most soil samples exhibit a pH value close to neutral, typically 6–7. The rainfall distribution has many peaks, with notable values at around 55 mm and 100 mm. The distribution indicates a significant disparity in rainfall among different samples. Nitrogen (N), phosphorus (P), potassium (K), and humidity exhibit bimodal distributions, suggesting the presence of two prominent values or conditions in the sample. Bimodal distributions in the dataset indicate the potential existence of separate groups or clusters, which could have significance in classification tasks. Variables such as temperature and pH have unimodal distributions, suggesting a constant and narrow range of values. The density plots offer a visual representation to observe the data's central tendencies, variability, and subgroups. A heatmap displaying the correlation between various features of the crop recommendation dataset is presented in Figure 1.4. Each feature pair has a correlation coefficient that ranges from −1 to 1. There is no correlation when the value is 0, a complete negative correlation when it is −1, and a fully positive correlation when it is 1. N is negatively correlated with P (−0.23) and K (−0.14), whereas it shows a slight positive correlation with humidity (0.19) and pH (0.10). On the other hand, P shows a strong positive correlation with K (0.74), a negative correlation with N (−0.23), and a slight negative correlation with temperature (−0.13), humidity (−0.12), and pH (−0.14). P has a strong positive correlation with P (0.74), a slightly negative correlation with temperature (−0.16), and a negative correlation with N (−0.14) and pH (−0.17). The temperature has a positive correlation with humidity (0.21) and a slightly negative correlation with P (−0.13) and K (−0.16). Rainfall only negatively correlates with pH, and other features don't have any significant relationship with rain. A slight negative correlation with P (−0.14), K (−0.17), humidity (−0.01), and rainfall (−0.11) is noticed by pH, whereas only a slight positive happens with

TABLE 1.3

Transformation of Original Labels into Encoded Categorical Values

Original	Rice	Maize	Jute	Cotton	Coconut	Papaya	Orange	Apple
Encoded	0	1	2	3	4	5	6	7
Original	Muskmelon	Watermelon	Grapes	Mango	Banana	Pomegranate	Lentil	Black gram
Encoded	8	9	10	11	12	13	14	15
Original	Mungbam	Moth bean	Pigeon peas	Kidney beans	Chickpea	Coffee		
Encoded	16	17	18	19	20	21		

N (0.10). A similar analysis goes for humidity. Any value less than 0.1 in magnitude indicates almost no correlation. , which is why these values are ignored in this analysis.

1.4.3 FEATURE ENGINEERING

After data preprocessing, the dataset is ready to be processed, and the feature engineering part comes. First, the names of all the crops presented in this dataset were encoded categorically, and Table 1.3 describes the encoding of the labels.

After the encoding, the values of the features were scaled. This step was taken to improve the model's ability to classify. Scaling gives the ML model more leverage to work better on unseen data. In this study, the StandardScaler technique was applied. After scaling the whole dataset, the minimum value was −3.85 and the maximum value was 4.474. Equation (1.1) was used to perform the whole procedure:

$$z_i = \frac{x_i - \mu_x}{\sigma_x} \tag{1.1}$$

Here,

- Original feature value of i th sample = x_i
- Standardized feature value = z_i
- Mean of the feature = μ_x
- Standard deviation = σ_x

After scaling, the input dataset was split into the training and testing parts. Training uses 80% of the data, while testing uses the remaining 20%.

1.4.4 MODEL TRAINING, EVALUATION, AND FINAL MODEL SELECTION

The model has been trained separately using nine distinct ML algorithms: LR (Logistic Regression), GNB (Gaussian Naive Bayes), SVM (Support Vector Machine), KNN (K-Nearest Neighbors), DT (Decision Tree), ET (Extra Trees), RF (Random Forest), XGB (Extreme Gradient Boosting), and Bagging classifier. Three different customized deep learning approaches – DCN (Deep and Cross Network), INN (Improved Neural Network), and ABNN (Attention-Based Neural Network) – have also been used separately to train the model. The training was done using 80% of the data from the dataset. Several hyperparameters have been used to tune the ML model to be the most accurate model. Following the completion of the training procedure, the performance of this ML model has been assessed using the remaining 20% of data from the dataset. Six standard metrics have been used, which are "Accuracy," "Precision," "Recall," "F1-score," "AUC–ROC," and "Log Loss," to evaluate the ML model, and their equations are given in the following sections.

1.4.4.1   Accuracy

This is a metric that quantifies the extent to which a model is right. The accuracy is defined as the proportion of accurately predicted instances for all classes in the dataset to the total number of instances:

$$Accuracy = \frac{TP + TN}{TP + TN + FP + FN} \tag{1.2}$$

1.4.4.2   Precision

This metric quantifies the correctness of positive instance predictions. The ratio of total true positives for all classes to the total predicted positives through the testing dataset:

$$Precision = \frac{TP}{TP + FP} \tag{1.3}$$

1.4.4.3   Recall

A metric that quantifies how well a model can identify all relevant positive instances. It is particularly important in situations when the omission of positive cases has significant consequences. This counts the number of successfully predicted positive occurrences:

$$Recall = \frac{TP}{TP + FN} \tag{1.4}$$

1.4.4.4   F1-Score

A metric that optimizes the trade-off between precision and recall. This measurement becomes critical when the goal is to minimize both accurately classified cases and incorrectly classified cases.

$$F1\,Score = 2 \times \frac{Precision \times Recall}{Precision + Recall} \tag{1.5}$$

1.4.4.5   AUC–ROC

AUC–ROC stands for the "Area under the Receiver Operating Characteristic Curve." It gives a full assessment of how well the classifier works across several classes. It shows the relationship between TPR and FPR. If the TPR is high and the FPR is low across all thresholds, the performance is good, and the curve moves closer to the top left corner. The score falls between 0 and 1. A score of 0.5 suggests random categorization, and less than 0.5 indicates worse performance, while a score of 1 represents accurate performance.

$$True\,Positive\,Rate\,(TPR) = \frac{TP}{TP + FN} \tag{1.6}$$

$$False\,Positive\,Rate\,(FPR) = \frac{FP}{FP + TN} \tag{1.7}$$

1.4.4.6   Log Loss

This metric quantifies the accuracy of a classifier by evaluating the predicted probability of different classes to the actual target values. A score of 0 for this metric indicates a perfect match between projected probabilities and actual labels, while an increase in this number indicates a decline in performance.

$$Log\,Loss = -\frac{1}{N}\sum_{i=1}^{N}\sum_{j=1}^{M}y_{ij}\log\left(p_{ij}\right) \tag{1.8}$$

where,

TP = True Positive
TN = True Negative
FP = False Positives
FN = False Negatives
N   = Total number of instances
M   = Number of classes
y_{ij} = 1 if instance i is a member of the class j, 0 otherwise
p_{ij} = Predicted probability for instance i be a member of the class j

These measures are essential for assessing the effectiveness of classification algorithms, each metric offers distinct insight into various aspects of model performance. The final model will be trained by one particular ML algorithm or deep learning approach that performs best according to these metrics.

1.5   PERFORMANCE EVALUATIONS AND RESULT ANALYSIS

As the dataset contains 2,200 instances (rows), 20% of it means 440 instances (rows). Based on these 440 instances, some confusion matrices have been generated. Figure 1.6 presents nine confusion matrices based on LR, GNB, SVM, KNN, DT, ET, RF, XGB, and Bagging classifier after the hyperparameter tuning. This multi-class confusion matrix involves understanding the performance of a classification model that predicts multiple classes. A multi-class confusion matrix is an extension of the binary confusion matrix, where each class has its row and column. The x and y labels in the confusion matrices contain numbers from 0 to 21. These 22 numbers (categorical encoded values) represent the 22 crop names present in the

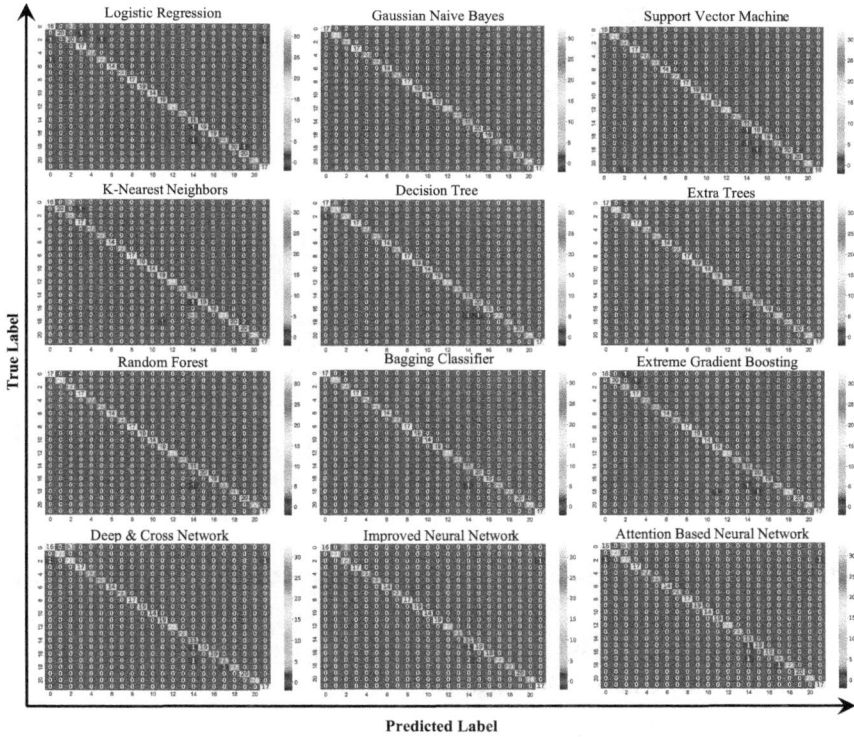

FIGURE 1.5 Comparison of confusion matrices for nine distinct algorithms used in model training.

dataset (target classes). Table 1.3 represents the crop name and corresponding categorical encoded value. In other words, it can be said that row and column numbers represent different crop names here. The y-axis displays the actual class, while the x-axis represents the predicted class. The diagonal elements of the confusion matrices represent the frequency of accurate predictions. For example, the first image of Figure 1.5 contains the LR algorithm's confusion matrix, "0" in the x-label, and the y-label indicates the name "Rice." This can be seen from the picture that the model indeed predicted 16 instances but 2 instances were misclassified. There is a value "1" in the 0th column of the 2nd and 5th row, which means that jute and papaya were misclassified as rice. That's why a total of 2 values are seen outside of the diagonal elements. The next column is for maize and 22 instances were accurately classified as maize; no misclassification happened here. The third column represents how many instances were classified as jute, and it can be seen that 20 instances were classified accurately and 3 instances were misclassified whose true class was rice but predicted as jute. A similar analysis goes for all columns and rows in this picture. After analyzing all the nine confusion matrices, it was concluded that GNB exhibits

Class	Prec.	Rec.	F1.	Sup.
0. Rice	1	0.89	0.94	19
1. Maize	1	1	1	21
2. Jute	0.92	1	0.96	23
3. Cotton	1	1	1	17
4. Coconut	1	1	1	27
5. Papaya	1	1	1	23
6. Orange	1	1	1	14
7. Apple	1	1	1	23
8. Musk melon	1	1	1	17
9. Watermelon	1	1	1	19
10. Grapes	1	1	1	14
11. Mango	1	1	1	19
12. Banana	1	1	1	21
13. Pomegranate	1	1	1	23
14. Lentil	1	1	1	11
15. BlackGram	1	1	1	20
16. Mung bean	1	1	1	19
17. Moth beans	1	1	1	24
18. Pigeon peas	1	1	1	23
19. Kidneybeans	1	1	1	20
20. Chickpea	1	1	1	26
21. Coffee	1	1	1	17
Accuracy			1	1
Macro Avg	1	1	1	440
Weighted Avg	1	1	1	440

(a)

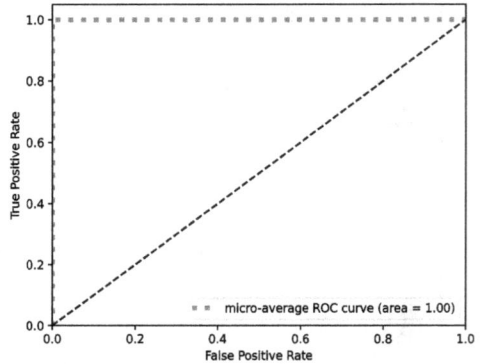

(b)

FIGURE 1.6   GNB classifier-based ML model's (a) classification report and (b) ROC curve.

the best results among all those algorithms. In total, 438 instances were predicted accurately and only 2 instances were misclassified.

Table 1.4 shows hyperparameters for nine distinct ML algorithms, the ranges of those parameters, and the ideal hyperparameter settings for achieving the best metric scores. The hyperparameter, penalty types, solver options, and regularization strength (C) were used for the LR algorithm. The best scores for accuracy, precision, recall, f1 score, AUC–ROC, and log loss were 97.05%, 97.15%, 97.05%, 97.04%, 99.97%, and 0.064, respectively, which had been obtained for penalty=l1, solver=saga, and C=10. While using the GNB algorithm, only one hyperparameter was used, and that was var_smoothing, and it was tuned for 1e−9. The best metric score found was 99.95%, 99.58%, 99.55%, 99.55%, 100%, and 0.0164 for accuracy, precision, recall, F1 score, AUC–ROC, and log loss, respectively. Next, the SVM algorithm was tuned with C, kernel, and gamma. This algorithm's best accuracy score was 97.95% for hyperparameter C=10, kernel=rbf, and gamma=scale.

The KNN algorithm gave the highest accuracy score of 97.50% when tuned with n_neighbors=3, weights=distance, and metric=Manhattan. The best accuracy for DT was 98.64%, which was obtained from tuning the hyperparameter to criterion=gini, splitter=best, max depth=none, min smples_split=2, min_samples_leaf=1, and max_features=none. On the other hand, six hypermeters were set for ET, which were n_estimators to 200, criterion to gini, max depth to none, min_samples_split to 10sssss, min_samples_leaf to 1 and max_features to sqrt. By doing so, the ET algorithm gave an accuracy of 98.86% in selecting the right crop. Moreover, the RF algorithm was tuned with n_estimators of 50, criterion of gini, max_depth of none, min_samples_split of 5 and min_samples_leaf of 1. This ended up providing a 99.32% accuracy score. The best accuracy Bagging

TABLE 1.4

Performance Analysis of Various Algorithms Used in the Training Process along with Hyperparameter Tuning

Algorithm	Hyperparameter	Range	Best	Accuracy	Precision	Recall	F1 Score	AUC–ROC	Log Loss
LR	Penalty	l1, l2, none, elasticnet	l1	97.05%	97.15%	97.05%	97.04%	99.97%	0.069
	C	"0.001, 0.01, 0.1, 1, 10, 100"	10						
	solver	Liblinear, saga, lbfgs	saga						
GNB	**var_smoothing**	**1e-9, 1e-8, 1e-7**	**1e-09**	**99.55%**	**99.58%**	**99.55%**	**99.54%**	**100%**	**0.016**
SVM	C	0.1, 1, 10, 100	10	97.95%	98.20%	97.95%	97.96%	99.98%	0.167
	Kernel	Linear, ploy, rbf, sigmoid	rbf						
	Gamma	Scale, auto	scale						
KNN	n_neighbors	3, 5, 7, 9	3	97.50%	97.89%	97.50%	97.53%	99.04%	0.688
	Weights	Uniform, distance	distance						
	Eetric	Euclidean, Manhattan, Minkowski	manhattan						
DT	Criterion	Gini, entropy	gini	98.64%	98.68%	98.64%	98.63%	99.28%	0.491
	Splitter	Best, random	best						
	max_depth	None, 10, 20, 30	none						
	min_samples_split	2, 5, 10	2						
	min_samples_leaf	1, 2, 4	1						
	max_features	srqt, log2, none	none						
ET	n_estimators	50, 100, 200	200	98.86%	99.05%	98.86%	98.88%	100%	0.158
	criterion	gini, entropy	gini						
	max_depth	None, 10, 20, 30	none						
	min_samples_split	2, 5, 10	10						
	min_samples_leaf	1, 2, 4	1						
	max_features	sqrt, log2, none	sqrt						

(Continued)

TABLE 1.4 (Continued)
Performance Analysis of Various Algorithms Used in the Training Process along with Hyperparameter Tuning

Algorithm	Hyperparameter	Hyperparameter Tuning		Accuracy	Precision	Recall	F1 Score	AUC–ROC	Log Loss
		Range	Best						
RF	n_estimators	50, 100, 200	50	99.32%	99.37%	99.32%	99.32%	99.99%	0.069
	Criterion	gini, entropy	gini						
	max_depth	None, 10, 20	None						
	min_samples_split	2, 5, 10	5						
	min_samples_leaf	1, 2, 4	1						
Bagging	n_estimators	10, 50, 100	100	99.32%	99.37%	99.32%	99.32%	100%	0.121
	max_samples	0.5, 1.0	1.0						
	max_features	0.5, 1.0	1.0						
	Bootstrap	True, false	True						
	bootstrap_features	True, false	True						
XGB	n_estimators	50, 100, 200, 300	200	98.86%	98.93%	98.86%	98.86%	100%	0.035
	learning_rate	0.01, 0.1, 0.2, 0.5	0.2						
	max_depth	3, 5, 7	3						
DCN	learning_rate	0.001, 0.01, 0.1	0.001	98.18%	98.27%	98.18%	98.19%	99.98%	0.057
	Epoch	200, 300, 400	200						
	Batch size	8, 16, 32	8						
	Dropout	0.1, 0.2, 0.3	0.1						
INN	learning_rate	0.002, 0.001	0.001	98.41%	98.62%	98.41%	98.43%	99.99%	0.984
	Epoch	200, 300, 400	300						
	Batch size	8, 16, 32	8						
	Dropout	0.1, 0.2, 0.3	0.1						
ABNN	learning_rate	0.0001, 0.0002	0.0001	98.18%	98.27%	98.18%	98.19%	99.98%	0.573
	Epoch	200, 300, 400	300						
	Batch size	4, 8, 16	8						
	Dropout	0.1, 0.2, 0.3	0.1						

classifier provided was similar to RF, and the hypermeters used to tune the Bagging classifier setting n_estimators to 100, max_samples and max_features both to 1, and bootstrap as well as bootstrap_features both making true. Lastly, XGB was tuned with three hyperparameters which were n_estimators of 200, learning_rate of 0.2, and max_depth, which ultimately offered an accuracy score of 98.86%.

Several DL approaches have been applied to observe how DL techniques perform on this tabular dataset. Hence, three DL methods – DCN, INN, and ABNN – were fined-tuned by the hyperparameters. DCN had utilized three cross-layers with five dense layers with 512, 256, 128, 64, and 32 units. Batch normalization and a dropout rate of 0.1 were used for better fitting and to reduce overfitting. The layers used ReLU activation and the final output layer had 22 units with softmax activation. DCN offered 98.18% accuracy scores when tested on the test dataset. On the other hand, ABNN has been trained with three fully connected dense layers (256, 128, 64 units) with ReLU activation and a dropout rate of 0.1. After the third dense layer, an attention mechanism with a dense layer (64 units), softmax activation, and an output layer of 22 units exhibited a 98.18% accuracy score, which is similar to the DCN model. Finally, INN has been utilized and INN provided the best accuracy score out of all these three DL approaches. The accuracy of INN-based DL model is 98.41%. Table 1.5 provides a brief description of the INN model.

TABLE 1.5
Improved Neural Network (INN) Model Summary

Layer (Type)	Output Shape	Param #
dense_12 (Dense)	(8, 512)	4,096
batch_normalization_6 (BatchNormalization)	(8, 512)	2,048
dropout_6 (Dropout)	(8, 512)	0
dense_13 (Dense)	(8, 256)	131,328
batch_normalization_7 (BatchNormalization)	(8, 256)	1,024
dropout_7 (Dropout)	(8, 256)	0
dense_14 (Dense)	(8, 128)	32,896
batch_normalization_8 (BatchNormalization)	(8, 128)	512
dropout_8 (Dropout)	(8, 128)	0
dense_15 (Dense)	(8, 64)	8,256
dense_16 (Dense)	(8, 32)	2,080
dense_17 (Dense)	(8, 22)	726

Total params: 182,968 (714.72 KB)
Trainable params: 181,174 (707.71 KB)
Non-trainable params: 1,792 (7.00 KB)
Optimizer params: 2 (12.00 B)

1.5.1   THE FINAL AI MODEL SELECTION

After observing the evaluation metrics in Table 1.4, it can be concluded that the GNB (Gaussian Naive Bayes) algorithm outperforms every ML algorithm used to train the model. On the other hand, INN (Improved Neural Network) performs better than the other two DL techniques. So by far, GNB and INN remain promising as ML algorithms and DL techniques. But when GNB is compared to INN, it is evident that GNB again outperforms INN in terms of accuracy, precision, recall, F1-score, and log loss. Moreover, when computation resources are considered, Table 1.6 can serve as a valuable reference. All the DL models took way more time than ML models in training and testing. So, the traditional ML models showed significant computational efficiency than advanced and complex DL models. There are other ML models except GNB and they showed better efficiency in computation than GNB. However, the computational efficiency difference between GNB and other ML models was not very high.

On the contrary, the accuracy of GNB is much better than other ML models and crop recommendation accuracy is important since one bad recommendation can be very destructive. That's why only GNB will be considered as an ML model worth comparing when compared with the customized fine-tuned DL model. Moreover, according to Table 1.6, GNB ranked second in time consumption to get trained. GNB took 0.0026 seconds to get trained, while INN took 165.15 seconds. Furthermore, GNB is almost 242 and 367 times quicker than INN while testing on the entire test

TABLE 1.6
Computational Time Required for All the Models

Model	Training Time (seconds)	Prediction Time on the Entire Test Set (seconds)	Prediction Time on Single Test Example (seconds)
LR	5.980017	0.000730	0.000135
GNB	0.002671	0.002603	0.000894
SVM	0.191121	0.042291	0.000289
KNN	0.002129	0.012350	0.000614
DT	0.008958	0.000385	0.000092
ET	0.454891	0.055812	0.010429
RF	0.195415	0.014336	0.003053
Bagging	0.823174	0.037857	0.008956
XGB	0.819125	0.009351	0.000618
DCN	130.705474	0.912402	0.396559
INN	165.158080	0.629289	0.328354
ABNN	126.813664	0.655981	0.386064

set and single test instance, respectively. So, the GNB classifier-based ML model is going to be used as the final model as it has the highest accuracy and good computational efficiency.

The classification report of the final ML model is described in Figure 1.6a, along with the Receiver Operating Characteristic (ROC) curve in Figure 1.6b for this multi-class classification. The displayed ROC curve is a "micro-average" ROC curve. Micro-averaging in multi-class classification involves combining the contributions of all classes to calculate the average metrics. This is achieved by globally calculating the metrics by counting the total number of TP, FN, and FP. The exhibited curve is a perfect ROC curve, demonstrating a perfect Area Under the Curve (AUC) value of 1.00. This indicates that the classifier performs optimally and accurately differentiates between the classes. The curve rises vertically from the point (0,0) to (0,1) and then runs horizontally from (0,1) to (1,1). This shape shows that the GNB classifier perfectly distinguishes between true and false positives. All these good scores may point to the "overfitting issue." However, all these metrics were generated or calculated based on the test instances (unseen data), as the entire dataset was partitioned into separate training and testing sets. Moreover, cross-validation scores are introduced in Figure 1.7, which shows that all individual fivefold scores are very high (close to 1), and the mean cross-validation score is also very high (99.5), suggesting that the model is consistently accurate across different subsets of the dataset. It indicates that the model is likely not overfitting to the training data but rather capturing genuine patterns that generalize well. All of these interpretations support the almost perfect classification performance metrics that were generated utilizing the test dataset. Due to the outstanding performance of the GNB classifier, this algorithm was selected to train the final ML model.

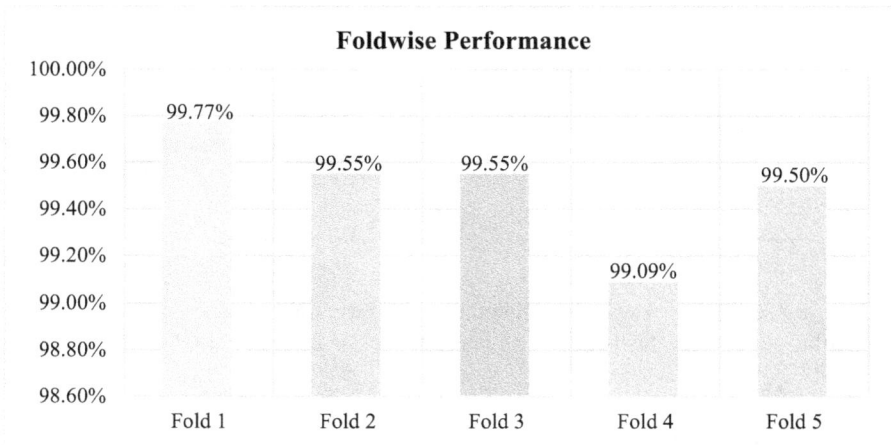

FIGURE 1.7 Foldwise cross-validation score of GNB-based model to check the overfitting issue.

1.5.2   Integrating XAI Tools with the GNB-Based Final ML Model

In this section, XAI methods that have been used to make this AI system more trustworthy and make the black-box model easier to understand will be explored thoroughly. The SHAP and LIME frameworks have already demonstrated their significance in the domain of XAI. SHAP and LIME precisely measure both the magnitude and sign of a feature's influence on a prediction. In my opinion, incorporating XAI analysis using both of these techniques should be a fundamental component of the ML pipeline. Different types of plots or pictures generated by SHAP and LIME are described in the following sections with their perfect interpretation. Crop names were encoded as numeric values, which is why all the target classes are represented as numeric values. In the upcoming sections, when a term such as "class 20" arrives, it means the crop name "Chickpea." Similarly, class 13 means "Pomegranate," and class 7 means "Apple," and the details can be found in Table 1.3.

1.5.3   SHAP Force Plot

The SHAP force plot indicates the features that exerted the greatest influence on the model's prediction for a particular observation. This local feature contribution is determined based on the SHAP values. It is highly advantageous for profoundly comprehending a specific case of prediction instance.

Code

```
# Assuming x _ test and explainer are already defined
# Choose the index of the instance you want to explain
chosen _ case = x _ test[[20]]
# Find the SHAP values for that specific case
shp _ vals _ case = explainer.shap _ values(chosen _ case)
# Flatting to 1D
shap _ vals _ case _ 1d = shap _ vals _ case[1].flatten()
# Plot the SHAP explanation for the chosen instance
force _ plot = shap.force _ plot(explainer.expected _ value[1], shap _
    vals _ case _ 1d,  chosen _ case,  feature _ names  =  original _
    feature _ names)
```

Figure 1.8a provides visual evidence of the prediction influence of features for a particular instance (x_test[[20]]) with respect to class 20 and it can be interpreted as follows:

- In Figure 1.8a, model's predicted probability for a particular instance, $f(x) = -2.59$, where the class was "1" (Rice). So it predicted the probability of an instance (x_test[[20]]) to be rice.
- Base value is the expected value in the absence of any known features which is 0.2812 in the chosen case. This value is obtained by determining

higher ⇄ lower
f(x)
-3.719   -2**-2.59**   -1.719   -0.7188   base value 0.2812   1.281   2.281

Nitrogen = 0.518   potassium = -0.2573   humidity = 0.3501 rainfall = 1.121 ph = 0.6959 temperature = -0.2323

(a)

higher ⇄ lower
f(x)
-2.90 719   -1.719   -0.7188   base value 0.2812   1.281

Nitrogen = -0.3939   potassium = -0.1398 humidity = 1.031 temperature = -0.1248 ph = 0.6896

(b)

higher ⇄ lower
f(x)
-1.719   base value 0.2812   2.281   4.281   6.281 **6.89**   8.281

potassium = -0.561 humidity = -0.2454 temperature = 1.422 ph = 0.4376 phosphorus = 0.3489   rainfall = -0.6157

(c)

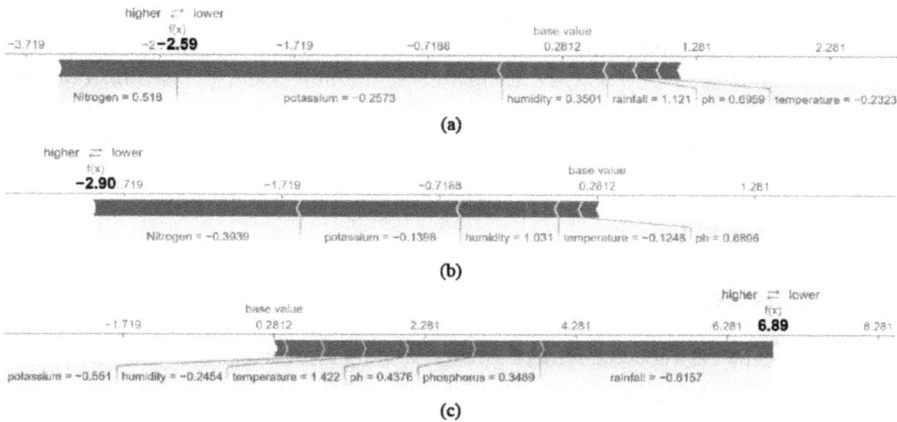

FIGURE 1.8   Force plot for (a) class "20," (b) class "13," and (c) class "21" (true class).

the mean of the model's outcomes over the training dataset and functions as a point of reference.

- The numerical values displayed below the plot arrows represent the feature's value of that particular instance. During the model training, the accuracy was increased by scaling the input dataset. Therefore, these are the scaled values. As shown in Figure 1.8a, the humidity was 0.3501 and potassium was −0.2573, among other features.
- Features that increased the prediction score are represented by red, while those that lowered it are represented by blue.
- The feature's impact on the prediction is directly proportional to the size of the arrow.
- According to Figure 1.8a, nitrogen has the maximum positive contribution having a feature value of 0.518, whereas all other features, potassium, humidity, rainfall, pH, and temperature, have negative contributions with feature values of −0.2573, 0.3501, 1.121, 0.6959, and −0.2323, respectively.
- The combined effect of these features adjusts the base value to shift to the final prediction of −2.59.

The same analysis applies to Figure 1.8b and c. Figure 1.8b and c displays the model's prediction for classes "20" and "13." The reason why $f(x)$ has a significantly high negative value in Figure 1.8a and b is because these two classes are not the actual class. Despite having a small positive base value, all other features contributed greatly to reducing the model's predictive score $f(x)$. Figure 1.8a had only one attribute that contributed to a positive predictive value, while Figure 1.8b did not have any such feature. So, it is evident that these classes are not genuine. Now, by modifying a single line in the provided code, the correct class was assigned, which is "21" (chickpea) and then Figure 1.8c was found where $f(x) = 6.89$.

```
shap _ vals _ case _ 1d = shap _ vals _ case[21].flatten()
```

All the features are represented because they increase the model's prediction probability. Rainfall has the highest impact on models' decisions, followed by several features such as phosphorus, pH, temperature, humidity, and potassium. For true class, $f(x)$ will be a positive value which is quite larger than the base value. As "21" is the true class, no features are involved in pushing the score backward and $f(x) = 6.89$ where base value is 0.2812. These graphs disclose information regarding the model's decision-making process by showing which factors increase the predicted outcome and which ones decrease it.

1.5.4 SHAP WATERFALL PLOT

The Waterfall plot is a specific form of graphic used for local analysis, with the main purpose of predicting a single instance. In the previous section, SHAP Force plots that utilize horizontal bars or arrows to visually depict contributions, creating a dynamic and easily understandable presentation, have been discussed. Force plots are superior in terms of rapidly and intuitively comprehending the influence of characteristics. On the other hand, waterfall plots use vertical bars to graphically indicate the cumulative stages that lead to the final choice. This plot offers a detailed and step-by-step dissection of how a prediction is obtained from the baseline, making it ideal for in-depth observations.

Figure 1.9a provides information about the prediction of a particular instance for class "0" and this figure can be explained as follows:

1. The model's predicted probability value, $f(x) = -3.176$.
2. The base value of model, $E[f(x)] = 1.07$.
3. In this diagram, the numbers on the left side show the feature values and the arrows show how much each feature affected the prediction.
4. These contributions are measured by how far they move the model output from the expected value based on the given instance's all-feature value.
5. The highest negative contribution is made by rainfall, with a value of -2.54, followed by smaller negative contributions from nitrogen, humidity, phosphorus, and temperature, with a value of -1.03, -0.61, 0.24, and -0.01, respectively. Two positive contributions were identified for potassium and pH, with a contributing value of 0.14 and 0.04, respectively.

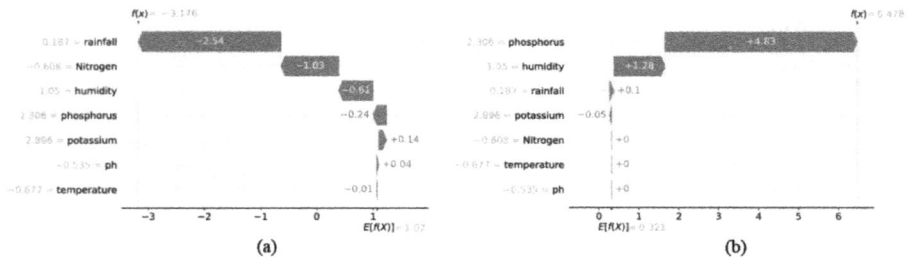

FIGURE 1.9 SHAP Waterfall plot of a particular instance (a) for false class and (b) for true class.

A similar analysis goes for Figure 1.9b for class "7." However, there are some differences between these two figures. For true class, $f(x)$ is always greater than $E[f(x)]$. Figure 1.9a is against false class and Figure 1.9b is for true class. As a result, among a total of seven features, we can see that five features are contributing toward a negative score of $f(x)$ in Figure 1.9a. Still, in Figure 1.9b, only one feature contributes to the negative score (even if the score is −0.05, a very small contribution) of $f(x)$. In contrast, the rest of the features contributed to the positive score, resulting in a high model predicted value, $f(x) = 6.478$. This means that Figure 1.9b represents the contribution of features for the true class.

1.5.5 SHAP BEESWARM PLOT

This powerful visualization technique effectively summarizes the SHAP values for several instances, facilitating the interpretation of ML model outputs. It thoroughly comprehends the overall distribution and impact of feature contributions across all instances in the dataset.

- Points on Figure 1.10 reflect SHAP values for features of every instance across the total dataset.
- On the y-axis, the features are arranged in descending order of importance, with the feature that is considered to be most important displayed at the top of the list. Based on the data presented in Figure 1.10, it is evident that "rainfall" holds the most importance, followed by "N" (nitrogen), "humidity," "K" (potassium), "P" (phosphorus), "pH," and "temperature" over the entire dataset.
- The x-axis shows the SHAP value, which is a measure of how each feature influences the model's output. The higher the SHAP number in magnitude, the more influential a feature is.
- The density of points represents the frequency at which specific SHAP values occur for this dataset. Concentrated groups of data points imply shared SHAP values throughout the dataset. Vertical jittering is applied to the points to prevent overlap and enhance the visibility of the distribution.

The impact of different features of all instances across the dataset is visualized in Figure 1.10. From the figure, it can be observed that great concentrations of points are located on the y-axis's left side, whereas only a few points are on the right side of the y-axis. All the low feature values of rainfall have been impacted negatively. However, higher feature values of rainfall sometimes have a positive impact and sometimes a negative impact. In the dataset, higher feature values with a positive impact are less in amount, whereas higher feature values with a negative impact are more in amount. This suggests that rainfall generally results in a decline in the expected prediction. Another essential feature that displays a diverse spectrum of SHAP values, encompassing both positive and negative aspects, is nitrogen. Greater nitrogen supply (shown by red points) generally enhances the predictions. In contrast, lower nitrogen levels (indicated by blue points) may either have a detrimental

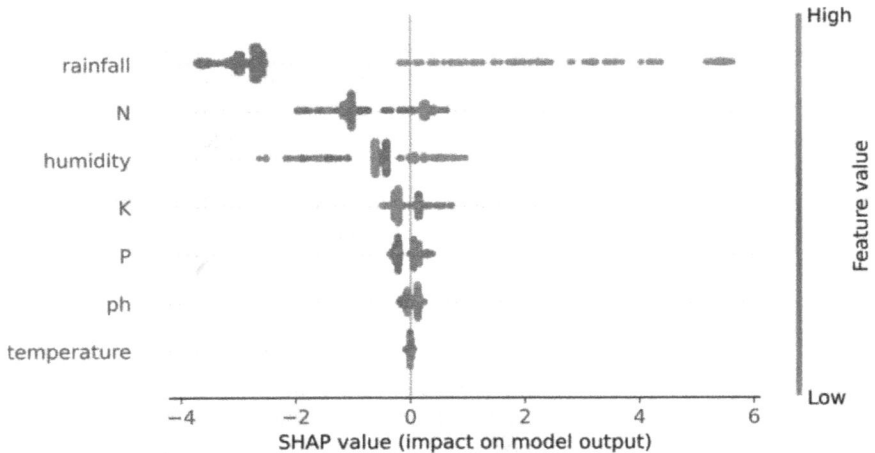

FIGURE 1.10   SHAP Beeswarm plot for all the instances in the dataset.

or less pronounced positive effect. It is clear from the diversity that nitrogen's effect on the model's predictions is context-based. In some circumstances, nitrogen might increase the prediction, but this is not always the case. Rather, most of the time, it decreases the prediction as the concentrations of dots are more on the left of the vertical axis representing SHAP values. The SHAP values for "Humidity" have a heterogeneous distribution, indicating its diverse influence on the model's predictions. The data points are distributed among positive and negative values, suggesting that humidity might positively or negatively impact the model's output, depending on its specific value for each occurrence. The absence of a distinct pattern indicates that humidity erratically affects the model's predictions.

Feature values of "potassium" (K) usually seem low. The blue dots are presented on both sides of the vertical SHAP value axis. However, they are more centralized around zero, but they still show some variation. This tells us that potassium levels are somewhat influential, though not as critical as rainfall, nitrogen, or humidity. Phosphorus has a moderate to low impact on the model output. The SHAP values for this feature show less variability and are mostly centered around zero. Whether high or low, feature values do not drastically affect the model's predictions. So, phosphorus has a smaller influence on the model's output compared to the top three features. The effect of pH on the model's results is moderate. Most of the SHAP values are near zero, indicating minimal impact. Both high and low pH values (red and blue dots) do not significantly alter the output. Among all the given features, temperature has the smallest effect on the model's forecast. The SHAP values for temperature are close to zero, indicating whether high or low feature value does not significantly influence the model's predictions. As far as the model's process for making choices is concerned, this feature is rather minor. The SHAP Beeswarm graphic reveals that rainfall, nitrogen, and humidity are the primary elements that significantly influence the model's predictions. Among these parameters, rainfall has the most substantial impact. Potassium and phosphorus exert a modest influence. Nonetheless, the results of the model are least affected by temperature and pH. This elaborate description enhances understanding of multiple factors' complex interactions and contributions to the model's predictions.

1.5.6 LIME CHARTS

ML models make predictions, and LIME is a specialized tool that tries to provide extensive explanations for each prediction by focusing on a small and localized context. This aims to clarify the reasoning behind a model's prediction for a given scenario. Spanning from simple linear regression to an advanced deep neural network, this approach is compatible with any ML model, making the LIME model agonistic. LIME's adaptability enables AI to generate explanations for various models without requiring knowledge of their internal mechanisms.

Code

```
# Assuming explainer, x _ test, and model GNB is already defined
# Explain the instance
explain  =  explaineer.explain _ instance(x _ test[3],  gnb.pre-
   dict _ proba, num _ features=7, top _ labels=1)
explain.show _ in _ notebook(show _ table=True)
```

There are four parts of the LIME explanation according to Figure 1.11:

i. The leftmost picture presents the probabilities of the predictions. The model predicts the instance as class "5" with a probability of 1.00 (100%), and there are no notable probabilities for any other classes.

ii. The second picture from the left illustrates the impact of seven distinct features. The numbers displayed on the horizontal bars indicate the relative significance of these characteristics:
   • *Humidity:* Improves prediction the most with a value of more than 0.83. High humidity is thus a strong indication that the incident should be classified as "5."

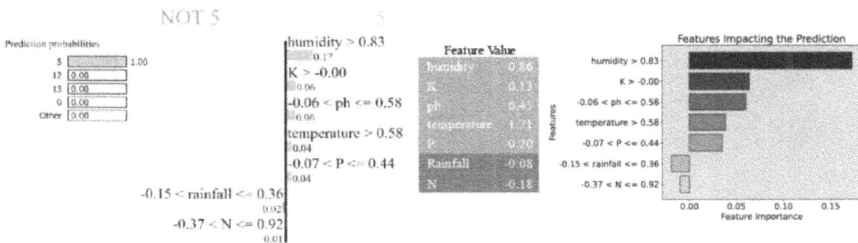

FIGURE 1.11   Explanation based on LIME.

- *K (potassium):* The instance is classified as "5" since a value more than −0.00 also favorably adds to the forecast.
- *pH:* The prediction is benefited by a range of −0.06 to 0.58, meaning a pH in this range raises the possibility of identifying the event as "5."
- *Phosphorus (P):* Phosphorus within this range positively affects the prediction, indicating that the classification as "5" is supported.
- *Temperature:* Higher temperatures increase the chance of identifying the incident as "5"; a value above 0.58 also favorably affects the prediction.
- *Rainfall:* A little positive effect indicates that rainfall falling between −0.15 and 0.36 is consistent with classifying the incident as "5."
- *Nitrogen (N):* The prediction benefits minimally from a range of −0.37 to 0.92, which helps to explain the "5" categorization.

iii. The real values of every feature are shown in the third figure from the left. As the dataset was scaled using StandardScaler, now all of the features have scaled values.

iv. The second figure from the left side and the first figure from the right side are identical. The latter figure is used for a better presentation.

Both SHAP and LIME have made significant contributions to XAI by shedding light on how ML models make decisions. The base of SHAP lies in Shapley values, which gives a rigorous and solid framework. However, LIME aims to generate multiple new samples by making small adjustments to the original data. This allows for more nuanced and detailed insights into how input changes affect the model's predictions. When it comes to making decisions in the real world, transparency and accountability are paramount, and both frameworks are necessary to bridge the gap between complex AI models and interpretability.

1.6   IMPLEMENTATION OF IoT TECHNOLOGY

Based on the performance evaluation outlined in Sections 1.4 and 1.5, it is clear that the GNB performed best among all the nine ML algorithms and three DL techniques that have been used in this research. That's why the final ML model has been trained using the GNB classifier. This trained ML model requires seven soil and environmental parameter information as input to recommend the best-suited crop. The parameters are nitrogen, potassium, phosphorus, temperature, humidity,

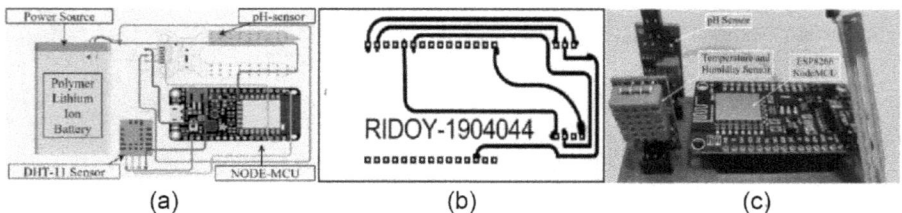

(a)   (b)   (c)

FIGURE 1.12   Schematic diagram of (a) circuit design, (b) PCB design, and (c) hardware setup of the circuit.

rainfall, and pH of a particular land where farmers want to grow their crops. So, instant data collection and monitoring support enhanced by available low-cost sensors and devices of IoT technology have been introduced in this study.

In this research, the farming field's temperature and humidity were monitored using a DHT-11 sensor. This sensor can detect both temperature and humidity through its thermistor and capacitive humidity sensor with $\pm 2°$ C and $\pm 5\%$ maximum expected deviation. On the other hand, the pH sensor has been used to measure the soil pH with $\pm 0.2\%$ maximum expected deviation. Both of these sensors give digital output which makes it easier for the sensors to interface with other devices. The connection between IoT devices, the corresponding printed circuit board (PCB) design, and the final hardware setup of IoT devices are illustrated in Figure 1.12.

Once the data collection is done, the next procedure is to upload the data to a server. To send the collected data to the server, NODEMCU ESP8266 was used. The IoT sensors are connected with ESP8266. The whole system is powered with the help of a rechargeable battery of 5V and a USB to type-B cable. ESP8266 is also connected to a Wi-Fi network. To connect with such a network, the name and password of that network are required. The Wi-Fi network can be a user's "mobile hotspot" or broadband-based network. After the connection and pairing are done, the IoT devices are ready to send data to the server. The "ThingSpeak" server was used to upload data, and from here, anyone with the ID and password of the server account can access and monitor the instant data that has been collected from the field. Users with any device can access the data whenever they want and after every 20 seconds, new data is written on the server. Figure 1.13 shows the real-time data on temperature, humidity, and pH that were gathered from the field. The final GNB classifier-based trained model along with the data on temperature, humidity, and pH are uploaded to a cloud platform. So far, only three types of information have been gathered, but the other four types of information are the value of nitrogen (N), phosphorous (P), potassium (K), and rainfall. Unfortunately, there are no cheap IoT sensors that can collect this information automatically. So, those are the pieces of information that need to be inserted manually in the user interface.

Figure 1.14 presents a website where users can insert these four types of essential information manually. The other information on temperature, humidity, and pH will be automatically updated on the website. Now after inserting all the information of the seven essential parameters of soil and the environment, the user should click on the "get recommendation" button. Once the button is pressed, this information will

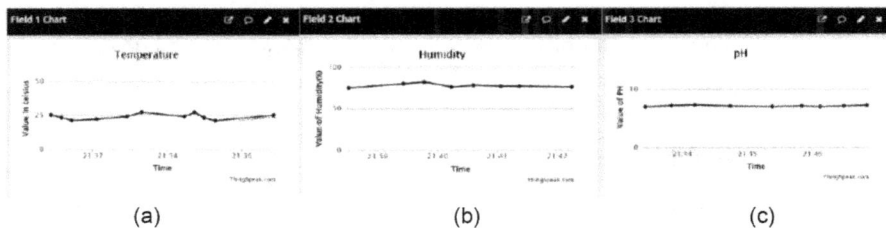

FIGURE 1.13   Real-time (a) temperature, (b) humidity, and (c) pH monitoring system.

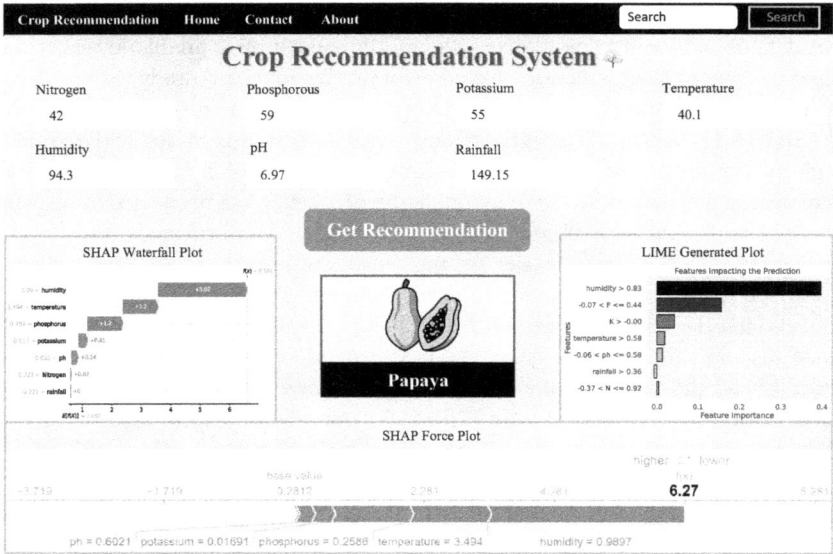

FIGURE 1.14 A user interface to enhance a data-driven crop selection scheme.

transfer to the cloud platform where our highly accurate trained model is present. The ML model will take the value of all the seven parameters of soil and the environment, process the information, and finally suggest a crop based on the received information. Moreover, the model will also generate several human-readable explanatory plots to provide logical reasoning for suggesting the corp. These XAI-based plots will demonstrate how the features have contributed in selecting a particular crop among the 22 crops. The suggested crop name and the plots will also travel through the API to the website, where users can see the recommended crop and the XAI plots. In Figure 1.14, a sample of the website has been provided. There, a user has provided information on nitrogen, phosphorous, potassium, and rainfall, whereas other information, such as on temperature, humidity, and pH, was already updated on the website as per the sensor's data. Users have access to change all the data before pressing the "get recommendation" button, and after pressing, it showed "Papaya" as the most suitable crop to cultivate according to the given information. There are also some XAI plots, revealing the influence of each soil and environmental parameter on the ML model in suggesting the name "Papaya." The plots revealed that humidity, phosphorous, potassium, and temperature had the highest influence in suggesting that crop name. In that condition of soil and environment, papaya would be best to cultivate.

1.7 DISCUSSION

Agriculture is an important element of every country's economy, yet it faces many challenges today like climate change, water scarcity, soil degradation, and so on. This hampers production and makes farmers vulnerable. Ensuring agricultural productivity maximization, resource management, and improving farmers' livelihood are growing global concerns. Choosing the best crop for harvesting according to the land and environmental context is the first step to address these concerns. In this research, a framework has been designed and developed to suggest to farmers which crop will be best for cultivation, and will also answer why the crop is suggested along with explanations presented as easily readable charts. By doing so, several key contributions are made, which will be discussed in the following sections as well as the limitations of this study.

1.7.1 CONTRIBUTION

This research developed a 99.55% accurate data-driven Gaussian Naive Bayes algorithm-based ML model that has been integrated with two XAI tools to recommend a crop and reveal the logic or reasonings behind the recommendation. Nine different ML algorithms and three customized fine-tuned DL models, which are based on neural networks, cross-connection, and attention mechanisms, have been applied separately to determine the most accurate AI model. Accuracy is crucial in crop recommendation since one wrong recommendation can cause a devastating socioeconomic impact on both the farmers and the country. A comparison between the ML and DL models has been performed. The results showed that the accuracy scores of the ML models were higher than DL models. Also, ML models were much more computationally efficient than DL models. The GNB classifier algorithm-based final AI model offered by this study takes 0.000894 seconds to test an instance which makes it computationally efficient. Such efficient ML models can perform well where resources are limited, making this AI model highly scalable for real-life implications.

SHAP and LIME are two prominent XAI methods that have been integrated into the final ML model to enhance the explainability of the complex ML model. Furthermore, XAI would increase trust by explaining the reasons behind a particular decision. As a result, the deployment of AI in the agricultural sector, where most individuals are not very familiar with AI systems, would become embraced and widely accepted. Furthermore, IoT technology has been used to collect some important information (temperature, humidity, and pH) directly from the farming field. These data are stored in a cloud server and are accessible whenever necessary. Hence, this study also provides a data monitoring system to continuously check the condition of the farming area. Moreover, a user-friendly interface has been created to serve farmers or agronomists of every category.

Many previous works have been performed on a similar topic which are presented in Table 1.7. Both ML and DL approaches have been taken to recommend a suitable crop for farmers to cultivate. For example, a majority voting (MV) approach was taken to recommend rice, ragi, gram, potato, and onion while having 94.78% accuracy in perfectly recommending. Moreover, other ML techniques such as RF (Random Forest), GB (Gradient Boosting) tree, and XGB (Extreme Gradient Boosting) have been used on various crops-related datasets to recommend a suitable crop in many studies [25, 45–47, 50]. In 2024, well-known DL techniques such as artificial neural networks (ANN) were utilized on 22 crops in two distinct studies

TABLE 1.7

Evaluation of This Research's Contribution Compared to Existing Literature

Reference	Year	Model	Number of Class (Crops)	Accuracy	IoT Technology	XAI SHAP	XAI LIME
Garanayak et al. [48]	2021	MV	5 crops	94.78	✗	✗	✗
Thilakarathne et al. [29]	2022	RF	22 crops	97.18	✗	✗	✗
Bhuyan et al. [49]	2022	GB tree	22 crops	99.11	✗	✗	✗
Musanase et al. [31]	2023	Neural Network	9 crops	97	✓	✗	✗
Senapaty et al. [34]	2023	MSVM-DAG-FFO	4 crops	97.3	✓	✗	✗
Ajoodha et al. [50]	2023	RF	10 crops	91.1	✗	✗	✗
Shingade et al. [51]	2023	RF	Multiple crops	95%	✗	✗	✗
Bhola et al. [26]	2024	ANN	22 crops	99.10	✗	✗	✗
Srinivasu et al. [52]	2024	RBF with SMO	22 crops	98	✗	✓	✗
Pasha et al. [53]	2024	ANN	22 crops	94%	✗	✗	✗
Dey et al. [54]	2024	XGB	21 crops	98.51	✗	✗	✗
Proposed Work	**2024**	**GNB**	**22 Crops**	**99.55**	✓	✓	✓

[26, 53] and their accuracy was 99.10% and 94%, respectively. In another study [52], radial basis functions neural network (RBF) and spider monkey optimization (SMO) model were used to classify 22 crops based on soil and environmental parameters in 2024. Only that particular study of Table 1.7 offered SHAP as an XAI tool, but there was no IoT technology or any user interface so that farmers and agronomists could benefit from the study directly. Most of the recent studies lack proper XAI implementation as per our investigation. Furthermore, a user-friendly interface was quite absent, but it is important as it prioritizes ease of use and accessibility, making advanced technologies more accessible to non-expert users like farmers. In this scenario, the 99.55% accuracy in crop recommendations with two prominent XAI tools integrated into the background of an interactive user interface offered in this study highlights the applicability in solving real-world problems.

1.7.2 LIMITATIONS

The limitation of this research is that IoT technology doesn't collect all the necessary data to recommend a crop. It collects only temperature, humidity, and pH from the targeted land where the farmer intends to grow crops. The other three essential characteristics which are nitrogen, phosphorus, and potassium are not collected automatically. Rather, they require manual insertions of value by the user. The nitrogen, phosphorus, and potassium sensors are quite costly, require calibration, and the chances of false results are quite high, whereas laboratory analysis of soil can be a good choice considering these issues. Farmers or agronomists are required to take samples of soil and go to the nearest soil testing laboratories to know the value of nitrogen, phosphorus, and potassium. After manually inserting these values, anyone can have a data-driven recommendation. However, the result of our study can also be generalized in various recommendation systems across different domains such as security [55], healthcare, and recommendations related to different environments [56–59].

1.8   CONCLUSION AND FUTURE WORK

This chapter presents a system to recommend crops that use IoT devices, XAI methodologies, and ML algorithms. The model was trained using many ML and customized fine-tuned DL approaches, and among them, GNB displayed the best performance. That's why the ML model used in this framework is based on the GNB classifier. A key component of this research is the introduction of XAI techniques. XAI makes the decision-making process understandable and trustworthy. This has a huge impact on farmers as they might be skeptical or hesitant to believe the AI-based system's suggestions. So, if a clear explanation of the method is provided for making decisions, they would find it less difficult to depend on the AI system's judgment. SHAP and LIME were used to interpret the ML model. Both of these packages explain which features contributed how much to the final predictions. The features that had a huge influence on a single prediction are crucial. These features should have the utmost attention and maximum resource allocation, because these are the significant ingredients or factors that distinguish a particular crop from others and force the model to select that particular crop. So, if these features vary to a threshold level, then crop production can be hampered. Moreover, from the SHAP Beeswarm plot, how the features contribute across the whole dataset was noted. Rainfall, nitrogen, humidity, and potassium play a huge value in distinguishing the crops, whereas temperature, pH, and phosphorus don't have any significant impact on model prediction. All these insights have been drawn from XAI technology. Finally, a highly accurate GNB classifier-based ML model with an accuracy of 99.5% was deployed to a farmer-friendly website from where farmers can get data-driven recommendations. After the collection of data in real time, IoT devices upload them to a server named "ThingSpeak." From that server, the website collects data automatically. The pH, temperature, and humidity fields are automatically filled out by the data collected through IoT devices. The rest of the fields, such as nitrogen, phosphorus, potassium, and rainfall, need to be filled out by hand manually. After entering all the data, any farmer on any scale can obtain customized recommendations and explanations.

In conclusion, all of this contributes in improving crop yield, optimizing resource allocation, and ultimately enhancing security in agricultural regions. Precision agriculture is the new cutting-edge technology of this era. This research contributes highly to the precision agriculture domain. Future researchers and developers can focus on the following:

  i. Developing and implementing cheap sensors for the collection of nitrogen, phosphorus, and potassium measurements from specific agricultural fields.

ii. Gathering more region/country-wise data of the same crop but different variants to train the ML model to make the recommendation system more customized and impactful.

iii. Including market demand and price analysis in addition to the proposed system so that farmers can maximize their profitability.

These technological advancements will offer a more sophisticated system for improving agricultural productivity and fostering a more resilient and efficient agricultural sector.

REFERENCES

[1] A. Kumar, S. Sarkar, and C. Pradhan, "Recommendation System for Crop Identification and Pest Control Technique in Agriculture," in *2019 International Conference on Communication and Signal Processing (ICCSP)*, Chennai, India: IEEE, 2019, pp. 0185–0189, https://doi.org/10.1109/ICCSP.2019.8698099.

[2] A. D. Tripathi, R. Mishra, K. K. Maurya, R. B. Singh, and D. W. Wilson, "Estimates for World Population and Global Food Availability for Global Health," in *The Role of Functional Food Security in Global Health*, Elsevier, 2019, pp. 3–24, https://doi.org/10.1016/B978-0-12-813148-0.00001-3.

[3] V. Reyes-García, *Routledge Handbook of Climate Change Impacts on Indigenous Peoples and Local Communities*, 1st ed., London: Routledge, 2023, https://doi.org/10.4324/9781003356837.

[4] A. Sharma, A. Jain, P. Gupta, and V. Chowdary, "Machine Learning Applications for Precision Agriculture: A Comprehensive Review," *IEEE Access*, vol. 9, pp. 4843–4873, 2021, https://doi.org/10.1109/ACCESS.2020.3048415.

[5] Y. Chen, J. Kuang, D. Cheng, J. Zheng, M. Gao, and A. Zhou, "AgriKG: An Agricultural Knowledge Graph and Its Applications," in *Database Systems for Advanced Applications*, G. Li, J. Yang, J. Gama, J. Natwichai, and Y. Tong, Eds., in Lecture Notes in Computer Science, vol. 11448, Cham: Springer International Publishing, 2019, pp. 533–537, https://doi.org/10.1007/978-3-030-18590-9_81.

[6] R. S. Pachade and A. Sharma, "Machine Learning for Weather-Specific Crop Recommendation," *International Journal of Health Sciences*, pp. 4527–4537, 2022, https://doi.org/10.53730/ijhs.v6nS8.13222.

[7] T. Van Klompenburg, A. Kassahun, and C. Catal, "Crop Yield Prediction Using Machine Learning: A Systematic Literature Review," *Computers and Electronics in Agriculture*, vol. 177, p. 105709, 2020, https://doi.org/10.1016/j.compag.2020.105709.

[8] T. Banavlikar, A. Mahir, M. Budukh, and S. Dhodapkar, "Crop Recommendation System Using Neural Networks," *International Research Journal of Engineering and Technology (IRJET)*, vol. 5, no. 5, 2018, pp. 1475–1480.

[9] P. P. Ray, "Internet of Things for Smart Agriculture: Technologies, Practices and Future Direction," *AIS*, vol. 9, no. 4, pp. 395–420, 2017, https://doi.org/10.3233/AIS-170440.

[10] M. Azrour, J. Mabrouki, A. Guezzaz, S. Benkirane, and H. Asri, "Implementation of Real-Time Water Quality Monitoring Based on Java and Internet of Things," in *Integrating Blockchain and Artificial Intelligence for Industry 4.0 Innovations*, S. Goundar and R. Anandan, Eds., in EAI/Springer Innovations in Communication and Computing, Cham: Springer International Publishing, 2024, pp. 133–143, https://doi.org/10.1007/978-3-031-35751-0_8.

[11] M. Mohy-Eddine, A. Guezzaz, S. Benkirane, and M. Azrour, "IoT-Enabled Smart Agriculture: Security Issues and Applications," in *Artificial Intelligence and Smart Environment: ICAISE'2022*, Springer, 2023, pp. 566–571.

[12] S. Dargaoui, M. Azrour, A. El Allaoui, A. Guezzaz, and S. Benkirane, "Authentication in Internet of Things: State of Art," in *Proceedings of the 6th International Conference on Networking, Intelligent Systems & Security*, ACM Digital Library, 2023, pp. 1–6.

[13] N. L. Tsakiridis, et al., "Versatile Internet of Things for Agriculture: An eXplainable AI Approach," in *Artificial Intelligence Applications and Innovations*, I. Maglogiannis, L. Iliadis, and E. Pimenidis, Eds, in IFIP Advances in Information and Communication Technology, vol. 584, Cham: Springer International Publishing, 2020, pp. 180–191, https://doi.org/10.1007/978-3-030-49186-4_16.

[14] A. A. Islam Ridoy, Md. A. Ismail Siddique, and O. Joyti, "A Machine Learning-Driven Crop Recommendation System with IoT Integration," in *2024 6th International Conference on Electrical Engineering and Information & Communication Technology (ICEEICT)*, 2024, pp. 812–817, https://doi.org/10.1109/ICEEICT62016.2024.10534479.

[15] S. Gawade, G. Rout, P. Kochar, V. Ahire, and T. Namboodiri, "Agroferdure: Intelligent Crop Recommendation System for Agriculture Crop Productivity Using Machine Learning Algorithm," in *2023 International Conference on Computer, Electronics & Electrical Engineering & Their Applications (IC2E3)*, Srinagar Garhwal, India: IEEE, 2023, pp. 1–9, https://doi.org/10.1109/IC2E357697.2023.10262476.

[16] A. Sharma, M. Bhargava, and A. V. Khanna, "AI-Farm: A Crop Recommendation System," in *2021 International Conference on Advances in Computing and Communications (ICACC)*, Kochi, Kakkanad, India: IEEE, 2021, pp. 1–7, https://doi.org/10.1109/ICACC–202152719.2021.9708104.

[17] N. N. Patil and M. A. M. Saiyyad, "Machine Learning Technique for Crop Recommendation in Agriculture Sector," *IJEAT*, vol. 9, no. 1, pp. 1359–1363, 2019, https://doi.org/10.35940/ijeat.A1171.109119.

[18] Ch. Rakesh D, V. Vardhan, B. B. Vasantha, and G. Sai Krishna, "Crop Recommendation and Prediction System," in *2023 9th International Conference on Advanced Computing and Communication Systems (ICACCS)*, Coimbatore, India: IEEE, 2023, pp. 1244–1248, https://doi.org/10.1109/ICACCS57279.2023.10113081.

[19] D. Modi, A. V. Sutagundar, V. Yalavigi, and A. Aravatagimath, "Crop Recommendation Using Machine Learning Algorithm," in *2021 5th International Conference on Information Systems and Computer Networks (ISCON)*, Mathura, India: IEEE, 2021, pp. 1–5, https://doi.org/10.1109/ISCON52037.2021.9702392.

[20] P. A, S. Chakraborty, A. Kumar, and O. R. Pooniwala, "Intelligent Crop Recommendation System using Machine Learning," in *2021 5th International Conference on Computing Methodologies and Communication (ICCMC)*, Erode, India: IEEE, 2021, pp. 843–848, https://doi.org/10.1109/ICCMC51019.2021.9418375.

[21] S. M. Pande, P. K. Ramesh, A. Anmol, B. R. Aishwarya, K. Rohilla, and K. Shaurya, "Crop Recommender System Using Machine Learning Approach," in *2021 5th International Conference on Computing Methodologies and Communication (ICCMC)*, Erode, India: IEEE, 2021, pp. 1066–1071, https://doi.org/10.1109/ICCMC51019.2021.9418351.

[22] P. Parameswari, N. Rajathi, and K. J. Harshanaa, "Machine Learning Approaches for Crop Recommendation," in *2021 International Conference on Advancements in Electrical, Electronics, Communication, Computing and Automation (ICAECA)*, Coimbatore, India: IEEE, 2021, pp. 1–5, https://doi.org/10.1109/ICAECA52838.2021.9675480.

[23] C. Sagana, et al., "Machine Learning-Based Crop Recommendations for Precision Farming to Maximize Crop Yields," in *2023 International Conference on Computer*

Communication and Informatics (ICCCI), Coimbatore, India: IEEE, 2023, pp. 1–5, https://doi.org/10.1109/ICCCI56745.2023.10128525.

[24] C. Hazman, A. Guezzaz, S. Benkirane, and M. Azrour, "lIDS-SIoEL: Intrusion Detection Framework for IoT-Based Smart Environments Security Using Ensemble Learning," *Cluster Computing*, pp. 1–15, 2022.

[25] M. Y. Shams, S. A. Gamel, and F. M. Talaat, "Enhancing Crop Recommendation Systems with Explainable Artificial Intelligence: A Study on Agricultural Decision-Making," *Neural Comput & Applic*, vol. 36, no. 11, pp. 5695–5714, 2024, https://doi.org/10.1007/s00521-023-09391-2.

[26] A. Bhola and P. Kumar, "Farm-Level Smart Crop Recommendation Framework Using Machine Learning," *Annals of Data Science*, 2024, https://doi.org/10.1007/s40745-024-00534-3.

[27] M. Hasan, et al., "Ensemble Machine Learning-Based Recommendation System for Effective Prediction of Suitable Agricultural Crop Cultivation," *Frontiers in Plant Science*, vol. 14, p. 1234555, 2023, https://doi.org/10.3389/fpls.2023.1234555.

[28] K. J. Kumar, N. P. K. Reddy, A. Vasu, and S. Sandosh, "A Proficient Approach in Crop Recommendation System Using Gradient Boosting Machine Technique," in *2023 3rd International Conference on Innovative Mechanisms for Industry Applications (ICIMIA)*, Bengaluru, India: IEEE, 2023, pp. 928–935, https://doi.org/10.1109/ICIMIA60377.2023.10425893.

[29] N. N. Thilakarathne, M. S. A. Bakar, P. E. Abas, and H. Yassin, "A Cloud Enabled Crop Recommendation Platform for Machine Learning-Driven Precision Farming," *Sensors*, vol. 22, no. 16, p. 6299, 2022, https://doi.org/10.3390/s22166299.

[30] Dolli, P. Rawat, M. Bajaj, S. Vats, and V. Sharma, "An Analysis of Crop Recommendation Systems Employing Diverse Machine Learning Methodologies," in *2023 International Conference on Device Intelligence, Computing and Communication Technologies, (DICCT)*, Dehradun, India: IEEE, 2023, pp. 619–624, https://doi.org/10.1109/DICCT56244.2023.10110085.

[31] C. Musanase, A. Vodacek, D. Hanyurwimfura, A. Uwitonze, and I. Kabandana, "Data-Driven Analysis and Machine Learning-Based Crop and Fertilizer Recommendation System for Revolutionizing Farming Practices," *Agriculture*, vol. 13, no. 11, p. 2141, 2023, https://doi.org/10.3390/agriculture13112141.

[32] V. D N and S. Choudhary, "An Artificial Intelligence Solution for Crop Recommendation," *IJEECS*, vol. 25, no. 3, p. 1688, 2022, https://doi.org/10.11591/ijeecs.v25.i3.

[33] S. Kiruthika and D. Karthika, "IOT-BASED Professional Crop Recommendation System Using a Weight-Based Long-Term Memory Approach," *Measurement: Sensors*, vol. 27, p. 100722, 2023, https://doi.org/10.1016/j.measen.2023.100722.

[34] M. K. Senapaty, A. Ray, and N. Padhy, "IoT-Enabled Soil Nutrient Analysis and Crop Recommendation Model for Precision Agriculture," *Computers*, vol. 12, no. 3, p. 61, 2023, https://doi.org/10.3390/computers12030061.

[35] E. M. B. M. Karunathilake, A. T. Le, S. Heo, Y. S. Chung, and S. Mansoor, "The Path to Smart Farming: Innovations and Opportunities in Precision Agriculture," *Agriculture*, vol. 13, no. 8, p. 1593, 2023, https://doi.org/10.3390/agriculture13081593.

[36] D. K. Kwaghtyo and C. I. Eke, "Smart Farming Prediction Models for Precision Agriculture: A Comprehensive Survey," *Artificial Intelligence Review*, vol. 56, no. 6, pp. 5729–5772, 2023, https://doi.org/10.1007/s10462-022-10266-6.

[37] S. A. Bhat and N.-F. Huang, "Big Data and AI Revolution in Precision Agriculture: Survey and Challenges," *IEEE Access*, vol. 9, pp. 110209–110222, 2021, https://doi.org/10.1109/ACCESS.2021.3102227.

[38] R. Akhter and S. A. Sofi, "Precision Agriculture Using IoT Data Analytics and Machine Learning," *Journal of King Saud University – Computer and Information Sciences*, vol. 34, no. 8, pp. 5602–5618, 2022, https://doi.org/10.1016/j.jksuci.2021.05.013.

[39] C. Murugamani, et al., "Machine Learning Technique for Precision Agriculture Applications in 5G-Based Internet of Things," *Wireless Communications and Mobile Computing*, vol. 2022, no. 1, p. 6534238, 2022, https://doi.org/10.1155/2022/6534238.

[40] W. Saeed and C. Omlin, "Explainable AI (XAI): A Systematic Meta-Survey of Current Challenges and Future Opportunities," *Knowledge-Based Systems*, vol. 263, p. 110273, 2023, https://doi.org/10.1016/j.knosys.2023.110273.

[41] R. Dwivedi, et al., "Explainable AI (XAI): Core Ideas, Techniques, and Solutions," *ACM Computing Surveys*, vol. 55, no. 9, pp. 194:1–194:33, 2023, https://doi.org/10.1145/3561048.

[42] S. M. Lundberg and S.-I. Lee, "A Unified Approach to Interpreting Model Predictions," in *Proceedings of the 31st International Conference on Neural Information Processing Systems*, in NIPS'17, Red Hook, NY: Curran Associates Inc, 2017, pp. 4768–4777.

[43] E. Štrumbelj and I. Kononenko, "Explaining Prediction Models and Individual Predictions with Feature Contributions," *Knowledge and Information Systems*, vol. 41, pp. 647–665, 2013, https://doi.org/10.1007/s10115-013-0679-x.

[44] L. S. Shapley, "17. A Value for n-Person Games," in *Contributions to the Theory of Games (AM-28), Volume II*, H. W. Kuhn and A. W. Tucker, Eds., Princeton University Press, 1953, pp. 307–318, https://doi.org/10.1515/9781400881970-018.

[45] M. T. Ribeiro, S. Singh, and C. Guestrin, "Anchors: High-Precision Model-Agnostic Explanations," *AAAI*, vol. 32, no. 1, 2018, https://doi.org/10.1609/aaai.v32i1.11491.

[46] J. Recio-García, B. Dí-az-Agudo, and V. Pino-Castilla, "CBR-LIME: A Case-Based Reasoning Approach to Provide Specific Local Interpretable Model-Agnostic Explanations," 2020, pp. 179–194, https://doi.org/10.1007/978-3-030-58342-2_12.

[47] "Crop Recommendation Dataset." Accessed: June 26, 2024   [Online].   Available: www.kaggle.com/datasets/shraddhasuman26/crop-recommendation-dataset/code.

[48] M. Garanayak, G. Sahu, S. N. Mohanty, and A. K. Jagadev, "Agricultural Recommendation System for Crops Using Different Machine Learning Regression Methods," *International Journal of Agricultural and Environmental Information Systems*, vol. 12, no. 1, pp. 1–20, 2021, https://doi.org/10.4018/IJAEIS.20210101.oa1.

[49] B. P. Bhuyan, R. Tomar, T. P. Singh, and A. R. Cherif, "Crop Type Prediction: A Statistical and Machine Learning Approach," *Sustainability*, vol. 15, no. 1, p. 481, 2022, https://doi.org/10.3390/su15010481.

[50] R. Ajoodha and T. O. Mufamadi, "Crop Recommendation Using Machine Learning Algorithms and Soil Attributes Data," in *Proceedings of 3rd International Conference on Artificial Intelligence: Advances and Applications*, G. Mathur, M. Bundele, A. Tripathi, and M. Paprzycki, Eds., in Algorithms for Intelligent Systems, Singapore: Springer Nature Singapore, 2023, pp. 31–41, https://doi.org/10.1007/978-981-19-7041-2_3.

[51] S. Shingade and R. Mudhalwadkar, "Sensor Information-Based Crop Recommendation System Using Machine Learning for the Fertile Regions of Maharashtra," *Concurrency and Computation: Practice and Experience*, vol. 35, 2023, https://doi.org/10.1002/cpe.7774.

[52] P. Naga Srinivasu, M. F. Ijaz, and M. Woźniak, "XAI-Driven Model for Crop Recommender System for Use in Precision Agriculture," *Computational Intelligence*, vol. 40, no. 1, p. e12629, 2024, https://doi.org/10.1111/coin.12629.

[53] S. N. Pasha, D. Ramesh, S. Mohmmad, and G. Deepak, "Automatically Suggesting Suitable Crop from Soil Samples Using Optimized Neural Networks," in *2024 3rd*

International Conference for Innovation in Technology (INOCON), Bangalore, India: IEEE, 2024, pp. 1–4, https://doi.org/10.1109/INOCON60754.2024.10511669.

[54] B. Dey, J. Ferdous, and R. Ahmed, "Machine Learning Based Recommendation of Agricultural and Horticultural Crop Farming in India Under the Regime of NPK, Soil pH and Three Climatic Variables," *Heliyon*, vol. 10, no. 3, 2024, https://doi.org/10.1016/j.heliyon.2024.e25112.

[55] M. Haque, D. Sonal, S. Ahmad, and K. Kumar, "Enhancing Security for Internet of Things Based System," 2023, pp. 869–878, https://doi.org/10.1007/978-981-99-3485-0_68.

[56] S. Jha, S. Routray, and S. Ahmad, "An Expert System-Based IoT System for Minimisation of Air Pollution in Developing Countries," *International Journal of Computer Applications in Technology*, vol. 68, p. 277, 2022, https://doi.org/10.1504/IJCAT.2022.124952.

[57] S. Ahmad and M. Yousuf Uddin, "An Intelligent Irrigation System and Prediction of Environmental Weather Based on Nano Electronics and Internet of Things Devices," *Journal of Nanoelectronics and Optoelectronics*, vol. 18, pp. 227–236, 2023, https://doi.org/10.1166/jno.2023.3382.

[58] M. Azrour, J. Mabrouki, A. Guezzaz, S. Ahmad, S. Khan, and S. Benkirane, Eds., *IoT, Machine Learning and Data Analytics for Smart Healthcare*, Boca Raton: CRC Press, 2024, https://doi.org/10.1201/9781003430735.

[59] A. Rajagopal, S. Jha, M. Khari, S. Ahmad, B. Alouffi, and A. Alharbi, "A Novel Approach in Prediction of Crop Production Using Recurrent Cuckoo Search Optimization Neural Networks," *Applied Sciences*, vol. 11, no. 21, Art. no. 21, 2021, https://doi.org/10.3390/app11219816.

2 Enhanced Date Palm Seed Classification Using Genetic Algorithms and XGBoost

Abdelaaziz Hessane, Mohamed Khalifa Boutahir, Ahmed El Youssefi, Yousef Farhaoui, and Badraddine Aghoutane

2.1 INTRODUCTION

Agriculture has significantly advanced through the integration of various scientific domains, with the classification of plant species playing a crucial role in enhancing agricultural productivity and biodiversity conservation [1]. This study focuses on the *Phoenix dactylifera* (date palm), a species of economic and cultural significance, particularly in arid regions [2, 3]. Traditional classification methods, which primarily rely on morphological and phenotypical characteristics [4–6], are often limited by their labor-intensive nature and susceptibility to human error, exacerbated by subtle phenotypic variations from environmental factors or genetic diversity.

The emergence of machine learning (ML) technologies offers advanced analytical capabilities for botanical classification, outperforming traditional methods by effectively detecting intricate patterns within complex datasets. This chapter proposes a novel ML-based classification approach for date palm seeds, optimizing the feature selection process using a Genetic Algorithm (GA). This method aims to determine a reduced yet effective feature set that enhances the accuracy and interpretability of classifications. The dual objectives of this research are to increase the computational efficiency of classification processes for practical application and to refine the feature set to improve model accuracy and interpretability. By aligning modern computational techniques with traditional botanical science, this approach seeks to revolutionize agricultural practices for *P. dactylifera* and potentially other species. The subsequent sections detail the materials, methods, experiments, and findings that support the viability and impact of this optimized classification approach.

DOI: 10.1201/9781003527664-2

2.2 RELATED WORK

Recent advancements in the field of botanical classification have been significantly propelled by machine learning. Researchers have utilized algorithms such as Support Vector Machine (SVM), K-Nearest Neighbor (KNN), and Multilayer Perceptron (MLP), focusing on features like leaf shape, size, color, and texture to achieve high accuracy in plant classification [7–11]. Techniques such as Pearson correlation and Information Gain have also been critical in feature selection, enhancing the efficacy of automated systems that analyze leaf images for consistent results [12]. While specific research on ML-based classification of date palms remains sparse, valuable insights have been drawn from the phenotypic and genetic diversity studies of date palm cultivars. These studies often employ methods like principal component analysis and multivariate analysis to explore phenotypic diversity and genetic polymorphism [4–6]. Moreover, the use of image analysis and machine learning has shown potential in distinguishing between date palm varieties by analyzing the morphological features of their fruits [13]. Notably, linear discriminant analysis has achieved an accuracy rate of 87.8% in differentiating *P. dactylifera* from other species using morphometric seed characteristics [14].

2.3 MATERIALS AND METHODS

This chapter's methodology is grounded in a rigorous and systematic machine learning approach to optimize the classification of *P. dactylifera* seeds. The process encompasses the preparation of the dataset, the application of machine learning models, and the utilization of a Genetic Algorithm for feature selection. Each step is designed to ensure the integrity and reproducibility of the results.

2.3.1 Dataset

This study utilizes an existing dataset as described by Gros-Balthazard et al. [14], which includes seeds from 13 *Phoenix* species. The dataset features comprehensive morphometric data on each seed, detailing both size and shape attributes. The size data includes measurements of length, width, thickness, and surface area, which quantitatively describe each seed's dimensions. Shape data, more complex, are derived from a geometric morphometric analysis using the elliptic Fourier transform (EFT). This analysis produced 64 shape features from 8 harmonics, captured from both lateral (VL) and dorsal (VD) views of the seeds, thus providing a detailed quantitative comparison of morphological differences among the seeds. The analysis framework transforms this into a binary classification task: Class 1 for *P. dactylifera* (571 samples) and Class 0 for non-*P. dactylifera* (1,054 samples), totaling 1,625 samples.

2.3.2 Machine Learning Models

Our research began with classical machine learning algorithms such as Decision Trees (DT), Support Vector Machines (SVM), and K-Nearest Neighbors (KNN) to establish baseline performance and understand the dataset's characteristics

[15, 16]. These widely used algorithms helped identify initial classification challenges and provided a reference point for evaluating more advanced techniques. As the study progressed, we shifted focus to ensemble methods, including Random Forest (RF), AdaBoost (ADA), Gradient Boosting Classifier (GBC), Extra Trees (ET), LightGBM (LGBM), and XGBoost (XGB), known for their robustness against overfitting and superior performance on complex datasets [17, 18]. These models leverage collective decision-making processes to enhance predictive accuracy and reliability. To fine-tune each model and maximize its performance, we applied a tenfold cross-validation technique, ensuring the algorithms' generalizability and preventing overfitting. This approach allowed for detailed performance comparisons and optimization of model parameters tailored to the dataset.

2.3.3 FEATURE SELECTION USING A GENETIC ALGORITHM

The research methodology's crux lies in applying a Genetic Algorithm (GA) for feature selection. The GA was configured with a fitness function defined by the accuracy of a candidate ML model. Through iterations of selection, crossover, and mutation, the GA searched for the most effective subset of features. The GA's performance was measured by the accuracy of the resulting model and the reduction in feature count. The aim was to identify a subset of features that could simplify the model without compromising – or potentially even improving – the classification accuracy. This research uses the simplest form of the genetic algorithm, as depicted in Algorithm 2.1 [19].

The algorithm iteratively improves a population of candidate solutions to an optimization problem. It starts by initializing a population of individuals (μ) and evaluating

---

Algorithm 2.1: The Simplest Form of a Genetic Algorithm

Input:

Output: The best individual found during the run, or, the best population found during the run.

```
1  ;
2    initialize;
3
4    while true) do
5
6
7  ;
8
9
end
```

their fitness ($F(t)$). In each iteration, the algorithm generates offspring (λ) through controlled recombination (), introduces mutations (), evaluates their fitness (), and selects the next generation () based on the combined fitness of parents and offspring, guided by the selection step size (). This process continues until a termination criterion, like a maximum number of iterations or a desired fitness level, is met. The algorithm outputs the best individual () and the best population () encountered during the run.

Traditional methods like principal component analysis PCA and filter-based techniques often assess features individually or in small groups, using statistical measures that may miss critical interactions vital for complex classification tasks [20]. In contrast, Genetic Algorithms (GAs) evaluate the collective impact of feature subsets, capturing essential nonlinear relationships that significantly affect model performance. The nonlinear optimization capabilities of GA are particularly suited for the complex nature of data in machine learning, enhancing model accuracy and robustness by leveraging detailed and interaction-rich datasets.

2.4 EXPERIMENTS AND RESULT ANALYSIS

All the experiments within this study were conducted using the Collaboratory tool of Google [21] and Python 3 programming language. The dataset was split into a training set (70%) and a testing set (30%). In addition, comprehensive experiments were designed to assess the effectiveness of the proposed approach. The first three experiments compared the performance of various machine learning algorithms on different feature sets derived from the date palm seeds data: metric features only, EFD-based shape features only, and a combination of metric and shape features. Figure 2.1 demonstrates the result of these experiments.

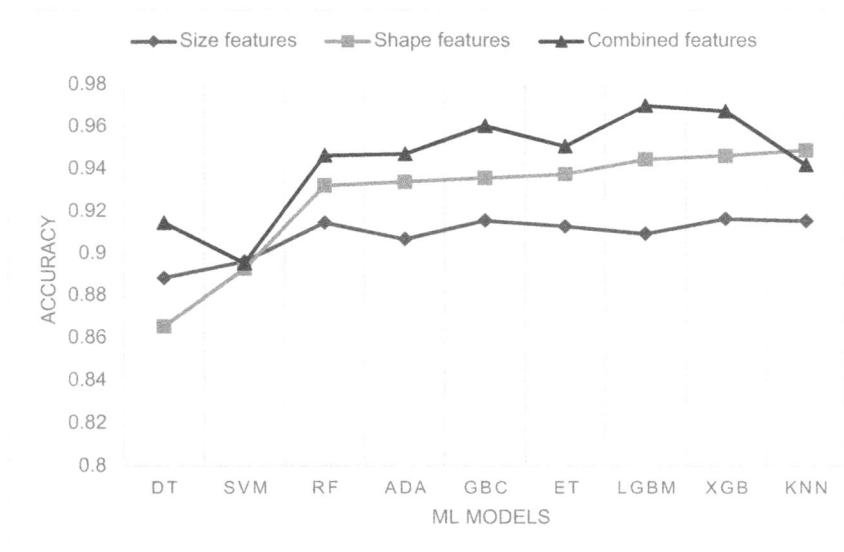

FIGURE 2.1   Classifier's accuracy per feature set.

TABLE 2.1

Genetic Algorithm Configuration

Parameter	Value
No. of Generation	30
Population size	20
Crossover probability (cxpb)	0.5
Mutation probability (mutbp)	0.3

TABLE 2.2

Impact of GA-Based Feature Selection on the Tested Classifier's Accuracy

Model	Accuracy	Features	No. of Features	Accuracy Using GA-Based Selected Features	Improvement
SVM	0.8962	Size features	4	0.877	−2.14%
DT	0.9147			0.9201	**0.59%**
RF	0.9464			0.9447	−0.18%
ADA	0.9472			0.9365	−1.13%
GBC	0.9604	Combined features	68	0.959	−0.15%
ET	0.9508			0.9467	−0.43%
LGBM	**0.9701**			0.9754	**0.55%**
XGB	0.9675			**0.9775**	**1.03%**
KNN	0.9490			0.8955	−5.64%

Ensemble methods like Random Forest, Gradient Boosting variations (GBC, ET), LightGBM, and XGBoost outperform basic classifiers like Decision Trees, SVM, and K-Nearest Neighbors by synthesizing multiple algorithms to handle complex datasets effectively. We further enhanced model performance using a Genetic Algorithm (GA) with LGBM's accuracy as the fitness function, optimizing feature selection across 30 generations. This optimization, especially with XGBoost, improved performance by 1.03%, reduced feature count from 68 to 32, and decreased computational load, thus increasing interpretability and training efficiency. Tables 2.1 and 2.2 depict the parameters of the GA, and the effect of using the GA-based set of features on the overall accuracy of the different classifiers, respectively. The parameters of the Genetic Algorithm, including population size, crossover probability, and mutation probability, were chosen empirically based on preliminary experiments to optimize performance.

Figure 2.2 illustrates the confusion matrix of the GA-optimized XGB model for date palm seed classification. Class 0 refers to the non-*P. dactylifera* seeds, and class

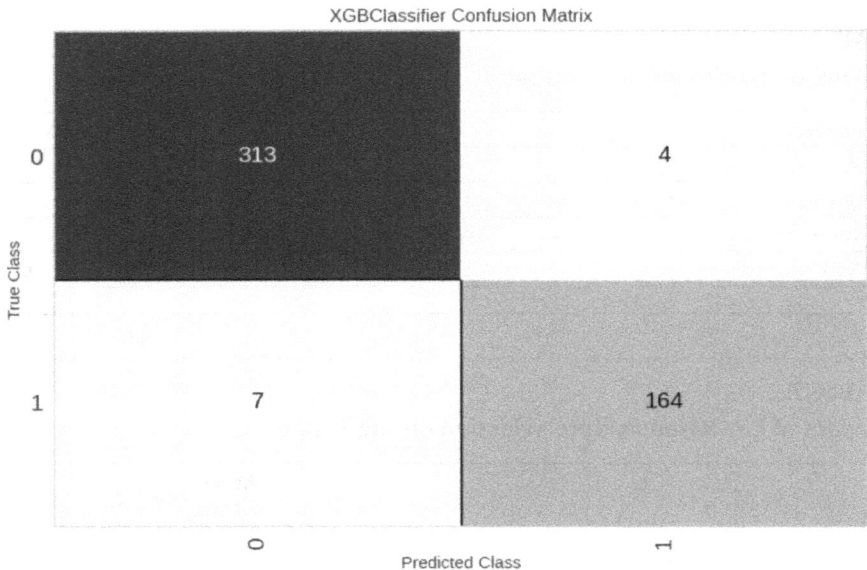

FIGURE 2.2   Optimized XGB classifier's confusion matrix.

1 refers to the *P. dactylifera* seeds. Leveraging a Genetic Algorithm (GA) for feature selection with the XGBoost model yielded promising results for date palm seed classification. This optimized approach achieved a remarkable 97.74% accuracy in differentiating *P. dactylifera* seeds (class 1) from non-*P. dactylifera* seeds (class 0). This success was achieved by selecting only 32 features, representing a 47% reduction from the original 68 feature set.

2.5   CONCLUSION AND FUTURE WORK

This chapter validates the effectiveness of using a Genetic Algorithm (GA) for feature selection in botanical classification tasks. By reducing features from 64 to 32, the GA not only preserved but also enhanced the classification accuracy, highlighting its utility in managing high-dimensional data. This method refined the dataset to its most crucial elements, increasing the model's accuracy, interpretability, and efficiency. Although GAs are computationally demanding due to their iterative and exhaustive nature, the trade-off between computational cost and improved model performance is pivotal, ensuring robust classification of date palm seeds. Looking forward, the study suggests exploring hybrid feature selection methods that combine the computational thriftiness of traditional techniques with the depth of GA's search capabilities, optimizing both performance and efficiency. Further research could extend this approach to a wider range of plant species, evaluate various feature selection algorithms, and integrate different types of data to enhance the precision of ML applications in agriculture. Such advancements could lead to more refined and interpretable models, promoting progress in precision agriculture.

REFERENCES

[1] N. Ben-Lhachemi, M. Benchrifa, S. Nasrdine, J. Mabrouki, M. Slaoui, and M. Ade Azrour, "Effect of IoT Integration in Agricultural Greenhouses," in *Technical and Technological Solutions Towards a Sustainable Society and Circular Economy*, J. Mabrouki and A. Mourade, Eds., Cham: Springer Nature Switzerland, 2024, pp. 435–445, https://doi.org/10.1007/978-3-031-56292-1_35.

[2] A. Hessane, A. El Youssefi, Y. Farhaoui, B. Aghoutane, and Y. Qaraai, "Artificial Intelligence Applications in Date Palm Cultivation and Production: A Scoping Review," in *Proceedings of the International Conference on Advanced Intelligent Systems for Sustainable Development (AI2SD'2022)*, Springer, 2023, pp. 230–239, https://doi.org/10.1007/978-3-031-26254-8_32.

[3] A. Hessane, M. Khalifa Boutahir, A. El Youssefi, Y. Farhaoui, and B. Aghoutane, "Empowering Date Palm Disease Management with Deep Learning: A Comparative Performance Analysis of Pretrained Models for Stage-wise White-Scale Disease Classification," *Data Metadata*, vol. 2, p. 102, 2023, https://doi.org/10.56294/dm2023102.

[4] K. Ennouri, R. Ben Ayed, S. Ercisli, S. Smaoui, M. Gouiaa, and M. A. Triki, "Variability Assessment in *Phoenix dactylifera* L. Accessions Based on Morphological Parameters and Analytical Methods," *Acta Physiologiae Plantarum*, vol. 40, no. 1, p. 5, 2018, https://doi.org/10.1007/s11738-017-2583-6.

[5] A. Simozrag, A. Chala, A. Djerouni, and M. Elmoncef Bentchikou, "Phenotypic Diversity of Date Palm Cultivars (*Phoenix dactylifera* L.) from Algeria," *Gayana Botanica*, vol. 73, no. 1, pp. 42–53, 2016, https://doi.org/10.4067/S0717-66432016000100006.

[6] A. Alahyane, et al., "Assessment of Phenotypic Diversity of Some Local Moroccan Date Palm Varieties and Clones (*Phoenix dactylifera* L.) from the Zagora Region, Southern Morocco," *Jordan Journal of Biological Sciences*, vol. 15, no. 4, pp. 671–678, 2022, https://doi.org/10.54319/jjbs/150416.

[7] V. Sharma, A. K. Mishra, and S. Paliwal, "Machine Learning Framework for Recognition and Classification of Plant Species," in *Proceedings of the 24th International Conference on Distributed Computing and Networking*, ACM, 2023, pp. 407–413, https://doi.org/10.1145/3571306.3571444.

[8] M. R. Sharan, A. A. Anil, N. Manohar, and B. R. Pushpa, "Classification of Medicinal Leaf by Using Canny Edge Detection and SVM Classifier," in *2022 International Conference on Futuristic Technologies (INCOFT)*, IEEE, 2022, pp. 1–6, https://doi.org/10.1109/INCOFT55651.2022.10094461.

[9] D. Barhate, S. Pathak, A. K. Dubey, and V. Nemade, "Cohort Study on Recognition of Plant Species Using Deep Learning Methods," *Journal of Physics: Conference Series*, vol. 2273, no. 1, p. 012006, 2022, https://doi.org/10.1088/1742-6596/2273/1/012006.

[10] B. K. Aman and V. Kumar, "Flower Leaf Image Classification Using Machine Learning Techniques," in *2022 Third International Conference on Intelligent Computing Instrumentation and Control Technologies (ICICICT)*, IEEE, 2022, pp. 553–558, https://doi.org/10.1109/ICICICT54557.2022.9917823.

[11] V. Jain and A. Yadav, "Analysis of Performance of Machine Learning Algorithms in Detection of Flowers," in *2021 Third International Conference on Intelligent Communication Technologies and Virtual Mobile Networks (ICICV)*, IEEE, 2021, pp. 706–709, https://doi.org/10.1109/ICICV50876.2021.9388599.

[12] P. Pungki, C. A. Sari, D. R. I. M. Setiadi, and E. H. Rachmawanto, "Classification of Plant Types Based on Leaf Image Using the Artificial Neural Network Method," in *2020 International Seminar on Application for Technology of Information and Communication (iSemantic)*, IEEE, 2020, pp. 67–72, https://doi.org/10.1109/iSemantic50169.2020.9234196.

[13] Y. Noutfia and E. Ropelewska, "Innovative Models Built Based on Image Textures Using Traditional Machine Learning Algorithms for Distinguishing Different Varieties of Moroccan Date Palm Fruit (*Phoenix dactylifera* L.)," *Agriculture*, vol. 13, no. 1, p. 26, 2022, https://doi.org/10.3390/agriculture13010026.

[14] M. Gros-Balthazard, C. Newton, S. Ivorra, M.-H. Pierre, J.-C. Pintaud, and J.-F. Terral, "The Domestication Syndrome in *Phoenix dactylifera* Seeds: Toward the Identification of Wild Date Palm Populations," *PLoS ONE*, vol. 11, no. 3, p. e0152394, 2016, https://doi.org/10.1371/journal.pone.0152394.

[15] A. Hessane, A. El Youssefi, Y. Farhaoui, and B. Aghoutane, "Toward a Stage-wise Classification of Date Palm White Scale Disease Using Features Extraction and Machine Learning Techniques," in *2022 International Conference on Intelligent Systems and Computer Vision (ISCV)*, IEEE, 2022, pp. 1–6, https://doi.org/10.1109/ISCV54655.2022.9806134.

[16] A. Hessane, A. El Youssefi, Y. Farhaoui, B. Aghoutane, and F. Amounas, "A Machine Learning Based Framework for a Stage-Wise Classification of Date Palm White Scale Disease," *Big Data Mining and Analytics*, 2023, https://doi.org/10.26599/BDMA.2022.9020022.

[17] M. K. Boutahir, A. Hessane, Y. Farhaoui, and M. Azrour, "An Effective Ensemble Learning Model to Predict Smart Grid Stability Using Genetic Algorithms," in *Proceedings of the 5th International Conference on Networking, Information Systems & Security (NISS 2022)*, Springer, 2023, pp. 129–137, https://doi.org/10.1007/978-3-031-25662-2_11.

[18] M. K. Boutahir, Y. Farhaoui, M. Azrour, A. El Allaoui, and E. M. Boumait, "A Novel OPT-GBoost Approach for Predicting Direct Normal Irradiance," in *Proceedings of the 1st International Conference on Smart Systems and Data Science (ICSSD 2022)*, Springer, 2023, pp. 343–350, https://doi.org/10.1007/978-3-031-35245-4_31.

[19] D. B. Fogel, T. Bäck, and Z. Michalewicz, *Evolutionary Computation. Vol. 1, Basic Algorithms and Operators*, Institute of Physics Publishing, 2000.

[20] M. K. Boutahir, A. Hessane, I. Lasri, S. Benchikh, Y. Farhaoui, and M. Azrour, "Dynamic Threshold Fine-Tuning in Anomaly Severity Classification for Enhanced Solar Power Optimization," *Data Metadata*, vol. 2, p. 94, 2023, https://doi.org/10.56294/dm202394.

[21] E. Bisong, "Google Colaboratory," in *Building Machine Learning and Deep Learning Models on Google Cloud Platform*, Berkeley, CA: Apress, 2019, pp. 59–64, https://doi.org/10.1007/978-1-4842-4470-8_7.

3 IoT-Based Precision Agriculture with Integrated Security for Smart Farming

Md. Alimul Haque, Sultan Ahmad, Deepa Sonal, Aasim Zafar, and Aleem Ali

3.1 INTRODUCTION

As the population continues to grow, we need to feed around 25% more mouths by 2050 than by 2010, a huge challenge taking into account the vast concerns of hunger that the world faces today. These figures are shocking, because over one-quarter of the world's population is malnourished [1, 2], and almost a billion people chronically suffer from hunger. Given the present scenario of starvation, feeding the new billions of mouths may be a massive demand in the future. Some people assume that the solution is actually to produce more food in order to deal with the crisis. The latest report published in the *Bioscience* has further enhanced that growth in the overall food supply in the bracket of 25–70% is scheduled between now and 2050 [3, 4]. But what if someone tells you that the food currently produced in the farms is adequate to feed 10 billion people?

Actually, the positive news is that many countries are now growing enough food, so it is important to determine how this food can be delivered while preserving its consistency. In view of this fact, one may argue that it is also important to increase food supply, but the actual subject of the food industry is to protect existing cultivation safely and effectually from various harmful animals and things [5, 6]. Nowadays, agricultural products/food protection challenges are attracting global attention and their protection traceability is one of the solutions embraced by all parts of the agri-food industry. State legislation and regulations for the promotion of food establishment have been implemented in many countries and territories systems of traceability and improving the safety of the food and agricultural products. The IoT-based framework for the agri-food supply chain will guarantee food protection from cropland to customer at any relation of production [7, 8].

The Internet of Things will become the foundation for Smart Computing's existence in the future [9–12]. The transformation of traditional technologies from home to workplace into "next-generation computing" plays a key role [13, 14]. "The Internet of Things" plays a central role in worldwide research and particularly in advanced

DOI: 10.1201/9781003527664-3

wireless communication [15]. IoT is laying the platform for the development of many items, like smart healthcare, smart housing, smart schools and technology today that impact people in and from the standard of the market. And it is used commercially in processing, shipping, agriculture and industry, and many other industries [16–18]. Agriculture is the most investigated component of IoT. When the global population is growing rapidly, it is really important to guarantee food security. Researchers first began to use information and communication technologies (ICT)-based technologies in this field, helpful in some areas but obviously not to fix the issue in the long term. Therefore, in cultivation, they now explore IoT as an alternative for ICT [19, 20].

Agricultural products include applications such as soil moisture control, temperature monitoring of the environment, humidity monitoring, supply chain management, and infrastructure management [21, 22]. Our proposed model is based on a generator of animal-friendly ultrasounds that does not harm animals either physically or biologically, or sounds humanly audible. This chapter provides information which can support IoT-based investigators and agricultural engineers to achieve the desired food security. The remainder of this chapter is arranged accordingly. Section 3.2 reviews the current literature. Section 3.3 details the proposed framework, including the components. Section 3.4 details the future scope in the product with the advancements in IoT. Finally, Section 3.5 concludes the work.

3.2 LITERATURE REVIEW

There is considerable work in the literature regarding the use of the Internet of Things (IoT) to deliver safety services for agriculture. ThingWorx [23] is an IoT system for handling transactions in connected worlds that provides the requisite protection and scalability. It provides rapid development assistance and technologies for on-site or data center deployments [12, 24]. This system is capable of providing the customer with the mandatory means of deploying smart IoT agro-based solutions with less effort. Fuse [25] is one of the precision agriculture solutions where farmers can access appropriate knowledge in order to take intelligent decisions, thus improving the efficiency and profitability of their firms. This approach is planned for the whole crop cycle process, planning, harvesting, and storing. Devanand et al. [26] provide a smart agriculture solution for the complete life cycle of the farming process and facilitates crop management via digital data-generated frameworks. Kaa [27] is an open-source IoT tool to operate Secure in farming, used for solutions such as mapping of resources, smart meters, and failure. It is built on a modular basis microdomain framework that enables the client to apply any kind of service necessary updates, enhancements, or integrations. Kaloxylos et al. and Mohy-Eddine et al. [28, 29], present a fully automated farm management system, Based on a specific module residing inside the cloud, the farm's management system functional architecture is discussed. In Baranwal et al. [30], an IoT-based system is presented that can analyze transmitted data is developed, checked, and evaluated. This system can be tracked and operated remotely and can be deployed in the agricultural domain for security purposes. The goal of this work is to provide real-term alerts and to resolve these problems without human interference on the basis of the data collected. In Jiang et al. [31], the authors have proposed a fully automated, large-scale, and wide-ranging tracking system based on WSN, for the tracking of honey bees and environmental data. Liu et al. [32], have introduced a ground-based WSN and Remote Sensing crop growth detection techniques. The communication system is based on the Zigbee modules and the managed optical sensor nodes. The proposed wireless connectivity is a point-to-point link with the controller between sensor nodes and gateway nodes. In another study [33], the authors addressed the areas of arable farming. IoT in farming can be considered as a next stage in the development. While other authors have discussed how IoT technologies can be helpful in managing livestock such as cows, goats, sheep, and other animals as well as monitoring their health [34]. Gorli and Yamini [35] described the roles of IoT in smart agriculture and the future of farming being dependent on these new IoT-based technologies. In a study by Nayyar and Puri [36], the authors used Arduino Super 2560 module and ESP8266 to display data on a computer screen; modern ESP32 microcontrollers were used as not every farmer has a PC, so real-time data for the Blynk smartphone app was used, which was quicker and more precise. Moreover, it also has a sleep mode to give the self-monitoring machine additional life. Human involvement in their product was still important. There are more expensive solutions as described by other researchers [37, 38] which can simplify the agriculture operation, but because of financial difficulties most farmers won't be able to do it. Our prototype only costs approximately 2,000 Indian rupees which could be more suited to farmers and which is also cheap and the code has a timer for sending readings for each trigger period in terms of redundancy so that the system does not receive duplicate inputs. Therefore, it provides higher-quality sensors, excellent code, routing algorithms, and a more sustainable architecture than the products suggested in by others [37, 39].

3.3   PROPOSED IoT MODEL

IoT is a low-energy application, trustworthy, and easy to use application-specific approach for addressing real-time issues. Sensors are information providers in the real network environment, and the actuators allow structures to reply to the sensed feedback. Relevant sensors in IoT are IR sensors, thermometer and infrared, optical monitor, smart watch, and IP camera. Farm crops are often destroyed by local animals such as buffaloes, pigs, goats, birds, and fire. For producers, this leads to big losses. Farmers cannot barricade entire fields or remain on the fields guarding them for 24 hours. So we have suggested an automated system of crop protection from animals and fire here. This proposed IoT model is going to enhance the features of the traditional inactive scarecrow to a smart versatile scarecrow that not only can frighten birds but also can repel wild animals and protect from fire.

3.3.1   WORKING PRINCIPLE

A traditional scarecrow sculpture feature will be modernized and improved in this proposed model by adding sensors and repeller devices that can detect animals in the range of croplands for farmers and as the model will be lashed with a fire sensor and alarm, it will detect fires in the cropland and buzz the alarm. We are connecting these sensor output signals with farmers' mobile phones using a GSM module so that

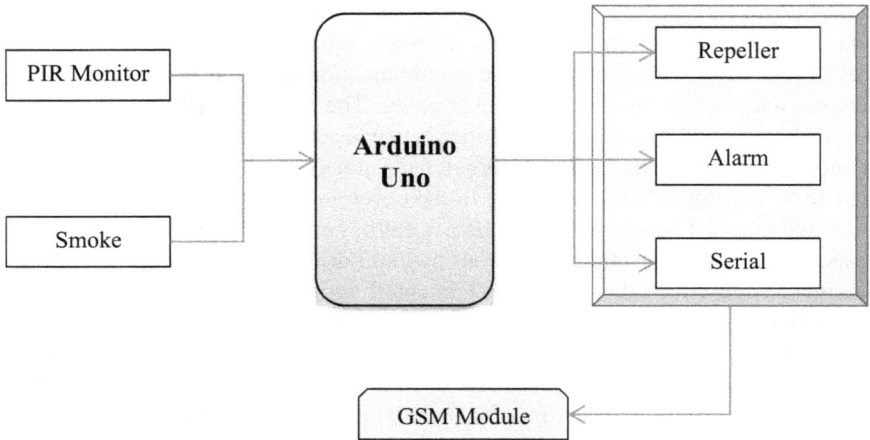

FIGURE 3.1   Proposed Smart Scarecrow model equipped with sensors, alarms, and solar cell.

when any movement is detected inside the farm in the farmer's absence, the related message can be sent to farmer, as shown in Figure 3.1.

3.4   ALGORITHM FOR PROPOSED MODEL

3.4.1   FOR ANIMAL DANGER

- Arduino UNO receives input from the PIR motion sensor when any animal enters the crop-land area of sensor range.
- As soon as the PIR sensor senses an animal, it will try to detect its height and Arduino will activate the output pin.
- At the output pin, the Repeller device will start producing an alarm sound of various high frequencies based on whether it is a small, medium, or large animal.
- At the same time, a message will be sent to the farmer's registered mobile phone using a GSM module.

3.4.2   FOR FIRE DANGER

- Arduino UNO receives input from the Smoke sensor when there is any fire in the crop-land area of sensor range.
- As soon as Smoke sensor senses the fire or smoke, Arduino will activate the output pin.
- At the output pin, the buzzer alarm will start producing a different alarm sound and a water motor will be switched on automatically.
- At the same time, a message will be sent to the farmer's registered mobile phone using a GSM module.

3.5   HARDWARE IMPLEMENTATION AND SPECIFICATIONS

3.5.1   PIR MOTION SENSOR

The Passive Infrared (PIR) sensor is a low-cost sensor that can recognize the location of animals or human beings. As seen in Figure 3.2, this sensor has three output pins for Vcc, Output, and Field. It can be used for any board such as Arduino, Raspberry, PIC, ARM, 8051, and so on, since the output pin is 3.3V TTL logic. It is possible to control the module from 4.5 V to 20 V, but usually 5 V is used.

3.5.2   ARDUINO

Arduino is an open-source microcontroller that can be quickly programmed, deleted, and reprogrammed at any time. Introduced in 2005, Arduino boards are based on Atmel microcontroller units (MCUs). On some more powerful devices, there is an additional microprocessor-based computer providing greater processing power and network communication. Arduino boards Uno, Nano, Mini 05, Mega 2560, Leonardo, Micro, Robot, Esplora, are based on Atmel MCUs with AVR architecture (Cvjetkovic & Matijevic, 2016):

- ATmega328 – Uno Nano and Mini 05
- ATmega2560 – Mega 2560
- ATmega32u4 – Leonardo, Micro, Robot, Esplora

FIGURE 3.2   PIR motion sensor.

The ATmega328/P provides the following features: 32 Kbytes of In-System Programmable Flash with Read While-Write capabilities, 1 Kbytes EEPROM, 2 Kbytes SRAM, 23 general-purpose I/O lines, 32 general-purpose working registers, real-time counter (RTC), three flexible Timer/Counters with comparable modes and PWM, one serial programmable USARTs, one byte-oriented 2-wire Serial Interface (I2C), 6-channel 10-bb Serial Interface (I2C). The Idle mode stops the CPU while allowing the SRAM, Timer/Counters, SPI port, and system interrupt to continue operation. The Power-down mode preserves the contents of the register but freezes the oscillator, removing all other features of the chip before the next interrupt or reset of the hardware. The asynchronous timer continues to run in power-save mode, enabling the user to retain a timer base while the remainder of the system is asleep. In order to reduce switching noise during ADC conversions, the ADC Noise Reduction mode interrupts the CPU and all I/O modules except the asynchronous timer and ADC. In Standby mode, when the majority of the system is asleep, the crystal/resonator oscillator works. Combined with low-power consumption, this facilitates very fast start-ups. Both the primary oscillator and the asynchronous timer begin to run in Prolonged Standby mode [40, 41]. An Arduino board is shown in Figure 3.3.

3.5.3 SMOKE SENSOR/FIRE SENSOR

This sensor is responsible for smoke detection by opto-electronic techniques and works on the theory of light scattering. It uses a beam of light emitted by a light-emitting diode (LED) SFH4551 from Osram (Regensburg, Bavaria, Germany)

FIGURE 3.3 An Arduino UNO.

1. Optical Chamber
2. PhotoDiode
3. Infra red LED

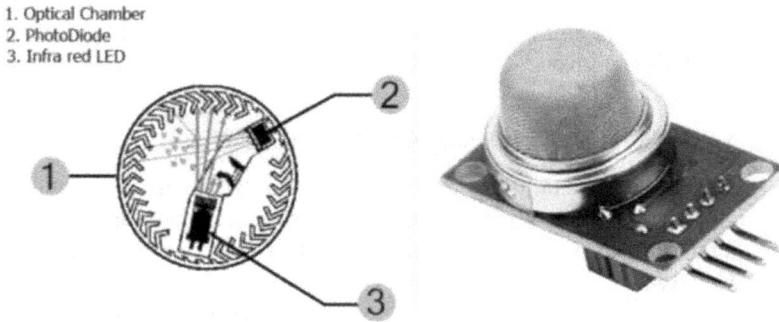

FIGURE 3.4   Smoke sensor chamber geometry and sensor image.

powered in a dark chamber that prevents the receiver (Osram's SFH2500 photodiode) from detecting the light passage due to the black material's absorption of light. As smoke particles join the light path in this chamber, light hits the particles and is mirrored on the microcontroller's photosensitive alerting unit. The chamber geometry is shown in Figure 3.4 [42].

3.5.4   REPELLER DEVICE

A repeller device consists of a low-power state-of-the-art Cortex ARM M0+ microprocessor that takes care of the output of frequencies and networking activities. One of the main features used is the unlicensed 2.4 GHz band IEEE 802.15.4 format, which enables small-size frames to be transmitted (about 11 bytes with very low-power consumption at a distance of 50 m). Along with LiPo batteries, the unit uses a solar panel, which makes it an autonomous energy device capable of operating even in times of partial or complete darkness. All the microcontroller pins are exposed, allowing a developer to attach sensors and actuators via digital and analog interfaces. We used a passive infrared (PIR) sensor to increase the energy efficiency of the system, which only activates the driver responsible for the generation of ultrasound and network contact when an animal is detected. The sound created by the system is 120 dB at a distance of about 1 m and a large band of 20–40 kHz that helps to adapt the device to the animal that is supposed to be repelled. We use a proxy program to receive, process, and archive the information retrieved by the computer, which gathers the animals' activity from the device and transmits it to the back-end machine [43]. Each repeller unit is installed in a watertight box which is mounted on a 4-m high pole.

3.6   GSM MODULE

A GSM or GPRS module is a chip or circuit used to link between a mobile device or a computer system and a GSM or GPRS system. A GSM/GPRS modem is a type of wireless modem designed for GSM and GPRS network communication. Similar to cell phones, a SIM (Subscriber Identity Module) card is required to enable network contact. They also have an IMEI (International Mobile Equipment Identity) number for their registration, equivalent to mobile phones.

3.7   FUTURE ENHANCEMENTS AND CHALLENGES

As seen in Figure 3.5, the built device can be operated by means of a mobile application created for iOS and Android. It enables system activation and deactivation, and provides additional functionality. You can create a smartphone application using Objective C and Java for iOS and Android, respectively. To produce push alerts, it is related to both Apple and Google application servers and notification systems, and the user will collect instant updates about any fire or animal attack accident in the field [42, 20].

However, the introduction of this technology presents its own unique challenges in India, particularly for the farmer who has limited land holdings and is located in a remote area without adequate Internet access and proper infrastructure where advanced monitoring systems are usually useless [31]. The high cost and sophistication of IoT devices can also be too expensive for farmers from modest backgrounds.

3.8   CONCLUSION

Agricultural industries would have to use revolutionary technology to gain the highly necessary edge to meet the burgeoning needs of the population. IoT agricultural applications can allow the industry to increase operating efficiency, reduce costs,

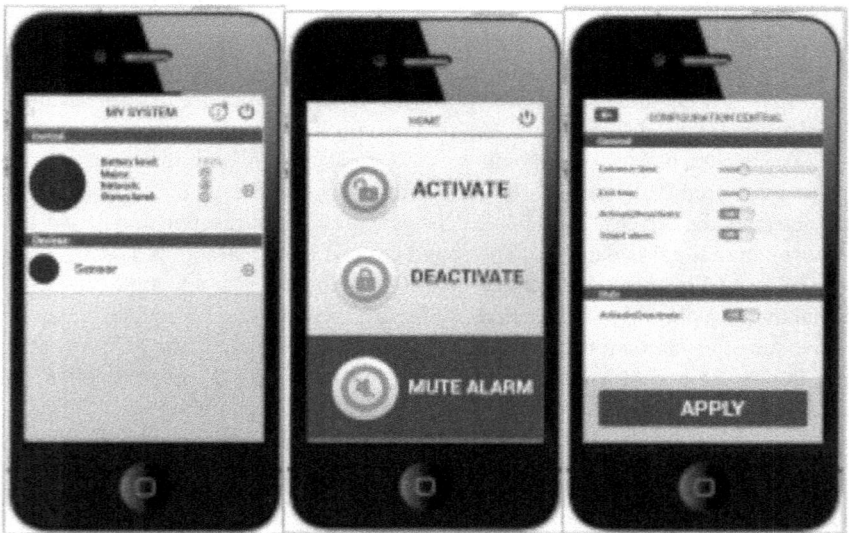

FIGURE 3.5   Mobile application for danger of fire and animal attacks.

reduce waste, and boost yield quality. IoT-based smart agriculture is a system which monitors irrigation activities and automates protection of crops in the agricultural field using sensors. Farmers can observe the condition of the farm from anywhere. The main purpose of this proposed model is to protect the crops in any condition and help farmers to control their land from anywhere.

ACKNOWLEDGMENT

We thank the Deanship of Scientific Research, Prince Sattam Bin Abdulaziz University, Alkharj, Saudi Arabia, for help and support by Project Number PSAU/2023/R/1444.

REFERENCES

[1] F. O. Adebayo and R. B. AbdusSalaam, "Jam Making and Packaging in Nigeria, Sub-Sahara Africa: A Review," *African Journal of Food Science and Technology*, vol. 10, no. 1, pp. 5–10, 2019.

[2] S. Nasrdine, M. Benchrifa, N. Ben-Lhachemi, J. Mabrouki, M. Slaoui, and M. Azrour, "New Design of an Inclined Solar Distiller for Freshwater Production: Experimental Study," in *Technical and Technological Solutions Towards a Sustainable Society and Circular Economy*, Springer, 2024, pp. 447–453.

[3] M. C. Hunter, R. G. Smith, M. E. Schipanski, L. W. Atwood, and D. A. Mortensen, "Agriculture in 2050: Recalibrating Targets for Sustainable Intensification," *Bioscience*, vol. 67, no. 4, pp. 386–391, 2017.

[4] M. Azrour, et al., "A Survey of Machine and Deep Learning Applications in the Assessment of Water Quality," in *Technical and Technological Solutions Towards a Sustainable Society and Circular Economy*, Springer, 2024, pp. 471–483.

[5] J. M. Mandyck and E. B. Schultz, *Food Foolish: The Hidden Connection Between Food Waste, Hunger and Climate Change*. Carrier Corporation, 2015.

[6] A. Haque, S. Haque, M. Rahman, K. Kumar, and S. Zeba, "Potential Applications of the Internet of Things in Sustainable Rural Development in India," in *Proceedings of Third International Conference on Sustainable Computing*, Springer, 2022, pp. 455–467.

[7] C. Costa, F. Antonucci, F. Pallottino, J. Aguzzi, D. Sarriá, and P. Menesatti, "A Review on Agri-Food Supply Chain Traceability by Means of RFID Technology," *Food and Bioprocess Technology*, vol. 6, no. 2, pp. 353–366, 2013.

[8] S. Haque, S. Zeba, Md. Alimul Haque, K. Kumar, and M. P. Ali Basha, "An IoT Model for Securing Examinations from Malpractices," *Materials Today: Proceedings*, 2021, https://doi.org/10.1016/j.matpr.2021.03.413.

[9] S. Dargaoui, et al., "IoT-Driven Smart Agriculture: Security Issues and Authentication Schemes Classification," in *Proceeding of the International Conference on Connected Objects and Artificial Intelligence (COCIA2024)*, Y. Mejdoub and A. Elamri, Eds., Cham: Springer Nature Switzerland, 2024, pp. 61–66, https://doi.org/10.1007/978-3-031-70411-6_10.

[10] N. Ben-Lhachemi, M. Benchrifa, S. Nasrdine, J. Mabrouki, M. Slaoui, and M. Ade Azrour, "Effect of IoT Integration in Agricultural Greenhouses," in *Technical and Technological Solutions Towards a Sustainable Society and Circular Economy*, J. Mabrouki and A. Mourade, Eds., Cham: Springer Nature Switzerland, 2024, pp. 435–445, https://doi.org/10.1007/978-3-031-56292-1_35.

[11] K. Bella, et al., "An Efficient Intrusion Detection System for IoT Security Using CNN Decision Forest," *PeerJ Computer Science*, vol. 10, p. e2290, 2024, https://doi.org/10.7717/peerj-cs.2290.

[12] S. Dargaoui, M. Azrour, A. Allaoui, A. Guezzaz, A. Alabdulatif, and A. Alnajim, "Internet of Things Authentication Protocols: Comparative Study," *CMC*, vol. 79, no. 1, pp. 65–91, 2024, https://doi.org/10.32604/cmc.2024.047625.

[13] W. Shafik and M. Azrour, "Building a Greener World: Harnessing the Power of IoT and Smart Devices for Sustainable Environment," in *Technical and Technological Solutions Towards a Sustainable Society and Circular Economy*, Springer, 2024, pp. 35–58.

[14] N. Ben-Lhachemi, M. Benchrifa, S. Nasrdine, J. Mabrouki, M. Slaoui, and M. Ade Azrour, "Effect of IoT Integration in Agricultural Greenhouses," in *Technical and Technological Solutions Towards a Sustainable Society and Circular Economy*, Springer, 2024, pp. 435–445.

[15] M. A. Haque, et al., "Internet of Things Enabled E-Learning System for Academic Achievement Among University Students," *E-Learning and Digital Media*, p. 20427530241280078, 2024.

[16] M. Mohy-Eddine, A. Guezzaz, S. Benkirane, and M. Azrour, "IoT-Enabled Smart Agriculture: Security Issues and Applications," in *Artificial Intelligence and Smart Environment: ICAISE'2022*, Springer, 2023, pp. 566–571.

[17] J. Mabrouki, et al., "Smart System for Monitoring and Controlling of Agricultural Production by the IoT," in *IoT and Smart Devices for Sustainable Environment*, Springer, 2022, pp. 103–115.

[18] N. Almrezeq, M. A. Haque, S. Haque, and A. A. A. El-Aziz, "Device Access Control and Key Exchange (DACK) Protocol for Internet of Things," *International Journal of Cloud Applications and Computing*, vol. 12, no. 1, pp. 1–14, 2022, https://doi.org/10.4018/IJCAC.297103.

[19] M. A. Haque, et al., "Sustainable and Efficient E-learning Internet of Things System Through Blockchain Technology," *E-Learning and Digital Media*, pp. 1–20, 2023, https://doi.org/10.1177/20427530231156711.

[20] V. Whig, B. Othman, A. Gehlot, M. A. Haque, S. Qamar, and J. Singh, "An Empirical Analysis of Artificial Intelligence (AI) as a Growth Engine for the Healthcare Sector," in *2022 2nd International Conference on Advance Computing and Innovative Technologies in Engineering (ICACITE)*, IEEE, 2022, pp. 2454–2457.

[21] M. Benchrifa, J. Mabrouki, M. Elouardi, and M. Azrour, "Studying the Effect of Integration Intelligent Dust Detection and Cleaning System on the Efficiency of Monocrystalline Photovoltaic Panels," in *Technical and Technological Solutions Towards a Sustainable Society and Circular Economy*, Springer, 2024, pp. 159–169.

[22] H. Hissou, H. Attou, S. Benkirane, A. Guezzaz, M. Azrour, and A. Beni-Hssane, "A Predicted Approach for Solar Radiation Using Multivariate Time Series," in *Advanced Technology for Smart Environment and Energy*, Springer, 2024, pp. 269–280.

[23] G. Suciu, C.-I. Istrate, and M.-C. Dițu, "Secure Smart Agriculture Monitoring Technique through Isolation," in *2019 Global IoT Summit (GIoTS)*, IEEE, 2019, pp. 1–5.

[24] S. Dargaoui, et al., "Internet-of-Things-Enabled Smart Agriculture: Security Enhancement Approaches," in *2024 4th International Conference on Innovative Research in Applied Science, Engineering and Technology (IRASET)*, IEEE, 2024, pp. 1–5.

[25] www.fusesmartfarming.com/agritechnica-2019-innovative-agco-solutions-on-display/.

[26] W. A. Devanand, R. D. Raghunath, A. S. Baliram, and K. Kazi, "Smart Agriculture System Using IoT," *International Journal of Innovative Research in Technology*, vol. 5, no. 10, pp. 480–483, 2019.

[27] B. Vogel, Y. Dong, B. Emruli, P. Davidsson, and R. Spalazzese, "What Is an Open IoT Platform? Insights from a Systematic Mapping Study," *Future Internet*, vol. 12, no. 4, p. 73, 2020.

[28] A. Kaloxylos, et al., "Farm Management Systems and the Future Internet Era," *Computers and Electronics in Agriculture*, vol. 89, pp. 130–144, 2012.

[29] M. Mohy-Eddine, A. Guezzaz, S. Benkirane, and M. Azrour, "IoT-Enabled Smart Agriculture: Security Issues and Applications," in *The International Conference on Artificial Intelligence and Smart Environment*, Springer, 2022, pp. 566–571.

[30] T. Baranwal and P. K. Pateriya, "Development of IoT Based Smart Security and Monitoring Devices for Agriculture," in *2016 6th International Conference-Cloud System and Big Data Engineering (Confluence)*, IEEE, 2016, pp. 597–602.

[31] J.-A. Jiang, et al., "A WSN-Based Automatic Monitoring System for the Foraging Behavior of Honey Bees and Environmental Factors of Beehives," *Computers and Electronics in Agriculture*, vol. 123, pp. 304–318, 2016.

[32] H. Liu, H. Sun, B. Mao, M. Li, M. Zhang, and Q. Zhang, "Development of a Crop Growth Detecting System," *IFAC-PapersOnLine*, vol. 49, no. 16, pp. 138–142, 2016.

[33] C. Verdouw, S. Wolfert, and B. Tekinerdogan, "Internet of Things in Agriculture," *CAB Reviews: Perspectives in Agriculture, Veterinary Science, Nutrition and Natural Resources*, vol. 11, no. 35, pp. 1–12, 2016.

[34] S. Jegadeesan and G. K. D. P. Venkatesan, "Smart Cow Health Monitoring, Farm Environmental Monitoring and Control System Using Wireless Sensor Networks," *International Journal of Advanced Engineering Technology*, vol. 334, p. 339, 2016.

[35] R. Gorli and G. Yamini, "Future of Smart Farming with Internet of Things," *Journal of Information Technology and Its Applications*, vol. 2, 2017.

[36] A. Nayyar and V. Puri, "Smart Farming: IoT Based Smart Sensors Agriculture Stick for Live Temperature and Moisture Monitoring Using Arduino, Cloud Computing & Solar Technology," in *Proceedings of the International Conference on Communication and Computing Systems (ICCCS-2016)*, 2016, pp. 9781315364094-121.

[37] J. Muangprathub, N. Boonnam, S. Kajornkasirat, N. Lekbangpong, A. Wanichsombat, and P. Nillaor, "IoT and Agriculture Data Analysis for Smart Farm," *Computers and Electronics in Agriculture*, vol. 156, pp. 467–474, 2019.

[38] J. Huuskonen and T. Oksanen, "Soil Sampling with Drones and Augmented Reality in Precision Agriculture," *Computers and Electronics in Agriculture*, vol. 154, pp. 25–35, 2018.

[39] T. Vineela, J. NagaHarini, C. Kiranmai, G. Harshitha, and B. AdiLakshmi, "IoT Based Agriculture Monitoring and Smart Irrigation System Using Raspberry Pi," *International Research Journal of Engineering and Technology*, vol. 5, pp. 1417–1420, 2018.

[40] D. Sonal, S. Haque, M. M. Nezami, and A. Balqarn, "An IoT-Based Model for Defending Against the Novel Coronavirus (COVID-19) Outbreak," *Solid State Technology*, vol. 63, no. 2, pp. 592–600, 2020.

[41] M. Haque, S. Haque, K. Kumar, M. Rahman, D. Sonal, and N. Almrezeq, "Security and Privacy in Internet of Things," in *International Conference on Emerging Technologies in Computer Engineering*, Springer, 2022, pp. 182–196.

[42] J. A. Luis, J. A. G. Galán, and J. A. Espigado, "Low Power Wireless Smoke Alarm System in Home Fires," *Sensors (Switzerland)*, vol. 15, no. 8, pp. 20717–20729, 2015, https://doi.org/10.3390/s150820717.

[43] S. Giordano, I. Seitanidis, M. Ojo, D. Adami, and F. Vignoli, "IoT Solutions for Crop Protection against Wild Animal Attacks," *2018 IEEE International Conference on Environmental Engineering, EE 2018 – Proceedings*, vol. 1, no. 710583, pp. 1–5, 2018, https://doi.org/10.1109/EE1.2018.8385275.

4 Dynamic Texture Classification Using ConvLSTM and Video Optical Flow Analysis
A Case Study

Manal Benzyane, Mourade Azrour,
Said Agoujil, and Azidine Guezzaz

4.1 INTRODUCTION

The importance of dynamic textures as valuable visual indicators for numerous video processing tasks, including facial expression identification [1, 2] and human action recognition [3], has led to a burgeoning interest in semantic video classification utilizing dynamic textures within the domain of computer vision in recent years. Despite the significant advancement, analysis of dynamic textures remains a challenging domain of research. The main focus of this investigation is on the mobility and appearance [4] of a dynamic texture.

Dynamic texture classification in agriculture involves the use of advanced algorithms to analyze and classify changing patterns in agricultural settings, such as crop growth and soil conditions. This field typically employs machine learning techniques and can significantly improve agricultural management practices by providing insights into soil texture and other dynamic factors affecting crop yields. The integration of methods like fuzzy C-means clustering has been noted as effective in classifying soil texture, which plays a vital role in sustainable agriculture management.

Dynamic texture classification involves categorizing video sequences based on their visual characteristics, such as clouds and flags among others [5]. In this context, the videos can be classified into categories such as Clouds/Steam, Fire, Flags, Trees, and Water. This classification utilizes deep learning techniques, including convolutional neural networks (CNNs) [6, 7], which integrate spatial and temporal features to improve accuracy in identifying and differentiating these dynamic textures [8].

The goal of the research is to categorize a specific video based on its dynamic texture sequences, which include non-rigid dynamic items like steam, flags, water, and fire [9]. Given that the shape and appearance of dynamic textures evolve over

DOI: 10.1201/9781003527664-4

time, this notion extends expanding upon the notion of self-similarity observed in stationary textures [10], dynamic textures are classified as belonging to the spatio-temporal domain.

In the realm of video processing and analysis, the integration of advanced techniques such as optical flow and Convolutional-Long-Short-Term Memory (ConvLSTM) networks represents a significant stride toward enhancing the accuracy and efficiency of video classification tasks [6, 7, 11, 12]. By leveraging optical flow in the preprocessing phase, followed by the application of ConvLSTM for video classification, this approach promises to unravel intricate temporal dynamics inherent in video datasets. Optical flow, a fundamental tool in computer vision, enables the tracking of pixel-level motion between consecutive frames, thereby providing crucial spatiotemporal cues essential for understanding dynamic content. Subsequently, the adoption of ConvLSTM networks, known for their adeptness in processing sequential data while retaining long-term dependencies, further enriches the classification process by effectively capturing both spatial and temporal features embedded within video sequences. This integration not only empowers the system to discern complex patterns within dynamic textures but also fosters a deeper comprehension of video content for accurate categorization. In this context, this study delves into the fusion of optical flow and ConvLSTM methodologies, elucidating their synergistic role in advancing video classification techniques and underscoring their potential to elevate the state of the art in multimedia analysis.

4.2 RELATED WORKS

The related work in dynamic texture classification encompasses several key approaches. First, a method employing local binary patterns (LBP) [13] for dynamic texture recognition is proposed, which extracts spatiotemporal features from sequences and demonstrates superior performance, emphasizing the significance of considering spatiotemporal information. Second, a two-stream convolutional neural network (CNN) [14] architecture is introduced for action recognition in videos, effectively capturing both appearance and motion cues through spatial and temporal streams, achieving state-of-the-art performance. Lastly, deep neural network architectures tailored for video classification [15] are presented, including methods for handling longer video sequences using convolutional temporal feature pooling architectures and recurrent neural networks with long short-term memory (LSTM) cells, showcasing significant improvements in classification accuracy, particularly for longer video sequences. These approaches collectively contribute to advancing dynamic texture classification and hold promise for applications in various domains such as video surveillance and sports analysis.

4.3 METHODOLOGY

In our case study, we conducted the project in one of three distinct phases. In phase one, we concentrated on preparing and classifying the dataset, making sure that there was a well-structured and tagged video collection available for the training and the testing the model. In addition to gathering the actual data, this phase also entailed necessary preprocessing tasks, such as standardization and enhancement, to boost the dataset's overall diversity and robustness. In the second phase, we applied optical flow techniques to the video data. This approach enabled us to capture motion patterns between frames, providing essential information on video dynamics. By calculating optical flow, we were able to represent object and camera motion in the scenes, enriching the features available for subsequent analysis. In the final phase, we used a long-term memory network (LSTM) to extract and categorize the features derived from the optical flow. This step involved training the LSTM to recognize and learn temporal dependencies within sequences of motion data, which proved vital for accurately understanding the context of the videos.

4.3.1 DATASET PREPARATION

The dataset we used consisted of 523 videos featuring dynamic textures, offering a rich variety of visual phenomena for analysis. These videos were meticulously categorized into five distinct classes: Fire, Clouds and Steam, Trees, Water, and Flags. Each category represents a different type of dynamic texture, providing a diverse range of challenges and characteristics for the classification task (Figure 4.1).

In the Fire category, we included videos that exhibited flickering flames and smoke patterns, capturing the chaotic yet mesmerizing movement of fire. The Clouds and Steam category contained clips showcasing the fluidity of clouds drifting and steam

FIGURE 4.1 Examples of the DynTex texture images.

TABLE 4.1

The Quantity of Videos Contained within the Dataset and Their Distribution across Different Categories

Categorize	Dataset
Cloud and Steam	109
Fire	104
Flags	104
Trees	101
Water	105
Total	**523**

billowing, emphasizing the soft transitions and variations in texture. The Trees category focused on the rustling branches and leaves swaying gently in the wind, representing organic movement. Water videos illustrated dynamic flows, splashes, and reflections, while the Flags category featured the fluttering movement of flags in the wind, characterized by their distinct shapes and colors.

Table 4.1 presents an overview of the total count of videos included in the dataset, along with their distribution across the different categories. This table not only highlights the diversity and volume of the dataset but also facilitates an understanding of the relative representation of each class, which is crucial for the training and evaluation of our video classification model [16]. By ensuring a balanced representation across categories, we aimed to enhance the robustness and generalizability of our findings.

4.3.2 OPTICAL FLOW

Optical flow [17] is the distribution of velocity components of light patterns in an image, that is, it is a field of velocity vectors associated with a sequence of images. In this sense, the optical flow entails a comprehensive velocity field, wherein every pixel within the image plane is linked to a distinct velocity vector. Thus, the optical flow serves as an approximation of the actual physical motion of the scene and allows a good description of the moving regions [16].

Figure 4.2 illustrates the application of optical flow in video analysis, demonstrating how this technique can effectively capture and visualize the motion between consecutive frames. The optical flow method computes the apparent motion of pixels across frames, providing valuable information about the direction and speed of moving objects within the video.

In the figure, we can observe various examples showcasing the optical flow vectors superimposed over the original video frames. These vectors indicate how specific regions of the image are moving at any given moment, with the length and direction of each vector representing the velocity and direction of motion, respectively. This visualization allows researchers to identify areas of interest, such as

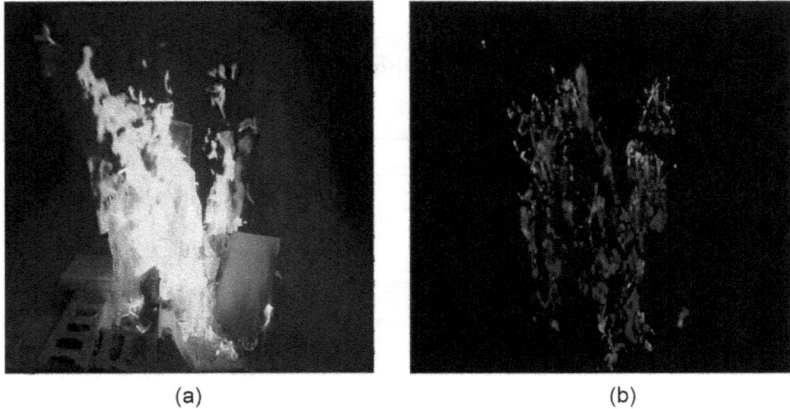

(a)   (b)

FIGURE 4.2   Optical flow application. (a) Original video. (b) Video optical flow.

rapid movements or changes in texture, and plays a crucial role in understanding the dynamics of the scene.

- Within a time increment dt, optical flow captures alterations in images resulting from motion.
- The optical flow field represents a velocity field illustrating the movement of points belonging to an object across a two-dimensional image in three-dimensional space.
- The optical flow should be robust to changes in lighting conditions or the movement of unimportant objects, such as shadows.
- If a moving source illuminates a fixed sphere, a non-zero optical flux will be noticed.
- No optical flow is produced by the smooth sphere revolving under steady illumination [18].

Figure 4.3. depicts samples of estimated optical flow, giving a comprehensive overview of how movement is portrayed in a stream of video images. This illustration presents various situations in which optical flow has been applied, highlighting the technique's capability to assess and quantitate the movement of various objects in a variety of different scenes.

The ability to identify and represent motion in a series of images is made possible by optical flow estimation techniques. Three major categories of techniques can be used to categorize optical flow calculation methods: differential-based, frequency-based, and correlation-based [19].

When using differential approaches, the initial hypothesis used to calculate the optical flow is that, for a brief period of time, the brightness between the many images in a series will be about constant. Therefore, using spatiotemporal derivatives of the original image, the speed is estimated [20, 21].

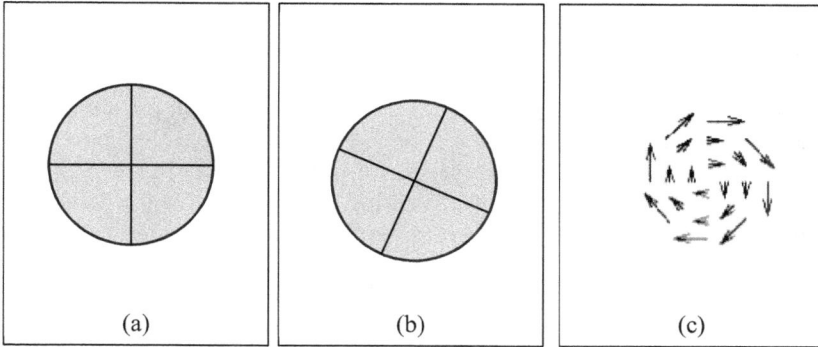

FIGURE 4.3 Example of optical flow estimation. (a) Frame at time t_0. (b) Frame at time $t_0 + 1$. (c) Estimated optical flux.

When measuring image motion, the initial assumption is that the local time-varying image region's intensity structures will remain roughly constant during the motion for at least a brief period of time. if $I(\vec{x}, t)$ is the image intensity function, then

$$I\left(\vec{x}, t\right) = I\left(\vec{x} + \delta\vec{x}, t + \delta t\right) \tag{4.1}$$

where $\vec{x} = (x, y)$ is the location vector, t is the time, $\delta\vec{x}$ is a quick trip to a particular area of the image, and δ_t is the time difference [22]. In other words, the displacement is viewed as insignificant for a short period of time. Using the Taylor series to expand the right side of Equation (4.1), we obtain:

$$I\left(\vec{x}, t\right) = I\left(\vec{x}, t\right) + \nabla I \cdot \delta\vec{x} + \partial t \, I_t + O^2 \tag{4.2}$$

where $I(\vec{x}, t)$ is the partial derivative and $\nabla I = (I_x, I_y)$ is the intensity spatial gradient vector consisting of the first-order partial derivatives of $I(\vec{x}, t)$, and O^2 are the higher order terms, which are assumed to be negligible. Subtracting the term $I\left(\vec{x}, t\right)$ from both sides, ignoring the higher order terms (O^2), and dividing the remaining terms by δt, we get

$$\nabla I \cdot \frac{\delta\vec{x}}{\partial t} + I_t = 0 \; \left(=\right) \nabla I \cdot \vec{v} + I_t = 0 \tag{4.3}$$

where $\vec{v} = \left(v_x, v_y\right)$ is the velocity field and v_x and v_y are, respectively, the velocities in the x and y directions The motion restriction equation is the name given to this equation. Equation (4.3) can be simplified as follows for 2D motion:

$$I_X \cdot v_x + I_y \cdot v_y + I_t = 0 \tag{4.4}$$

where I_x and I_y, respectively, stand for the x and y directional gradients. The velocity vectors v_x and v_y cannot be determined solely from this equation since there are two unknowns in the equation, which leads to an unlimited number of alternative solutions to the problem. To estimate the velocity components in this situation, new equations with unique constraints must be used. The fact that the optical flow is constant in the area that makes up a group, for instance, might be thought of as an extra constraint. Other instances include taking into account that the derivatives of the optical flow are constant, that the luminosity on a particular surface is constant regardless of the observing direction, or that only translation movements that are parallel to the picture plane are made [20, 22]. Optical flow refers to the displacement of an object between two successive frames within a sequence, resulting from the relative motion between the camera and the object. There are two varieties: dense optical flow and sparse optical flow. We work on dense optical flow.

4.3.3 CONVLSTM ARCHITECTURE

ConvLSTM represents a fusion of convolutional neural network (CNN) architecture with long short-term memory (LSTM) networks [16, 23–26]. In essence, it shares similarities with LSTM, yet integrates convolutional operations within layer transitions. The architectural layout of ConvLSTM is depicted in Figure 4.4.

ConvLSTM is particularly well-suited for processing images and videos characterized by temporal dependencies, allowing it to capture intricate spatiotemporal correlations within data. By leveraging inputs and past states from nearby cells in the grid, ConvLSTM forecasts the forthcoming state of a specified cell. This is facilitated by employing convolution operations during state-to-state and input-to-state transitions. Below are the fundamental ConvLSTM equations, where * denotes the convolution operator:

$$i_t = \sigma\left(W_{xt} * X_t + W_{ht} * H_{t-1} + W_{ci}C_{t-1} + b_i\right) \tag{4.5}$$

$$f_t = \sigma\left(W_{xf} * X_t + W_{hf} * H_{t-1} + W_{cf}C_{t-1} + b_f\right) \tag{4.6}$$

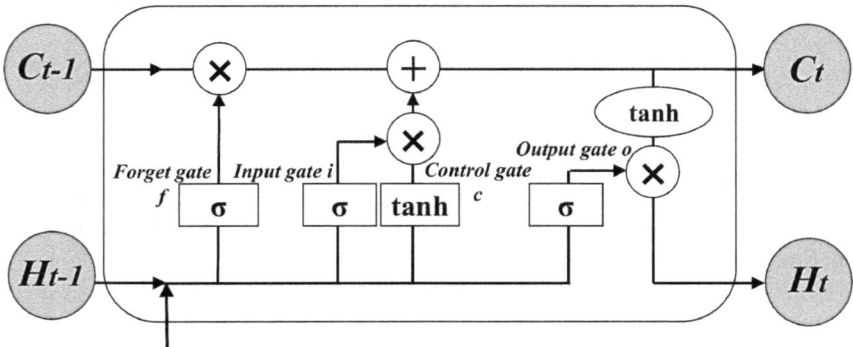

FIGURE 4.4 ConvLSTM cell architecture.

$$C_t = f_t C_{t-1} + i_t \tanh\left(W_{xc} * X_t + W_{hc} * H_{t-1} + b_f\right) \tag{4.7}$$

$$o_t = \sigma\left(W_{xo} * X_t + W_{ho} * H_t + W_{co}C_{t-1} + b_o\right) \tag{4.8}$$

$$H_t = o_t \tanh\left(C_t\right) \tag{4.9}$$

In the context of ConvLSTM, X_t represents the input to the cell, C_t denotes the cell output, and H_t signifies the hidden state of the cell. Additionally, i_t, f_t, and o_t correspond to the input gate, where the sigmoidal function σ is applied, and W denotes the convolution kernels [19].

4.4 CLASSIFICATION MODEL

In this study, we employed optical flow on our dataset and utilized ConvLSTM to extract features. To achieve this, we standardized the dimensions of each frame to 64×64 pixels and determined the number of frames constituting a sequence. Our approach involved extracting images from videos to generate the requisite data. The model sequentially processed individual images from the videos, extracting them one at a time. Upon extraction, the model provided a status indicator (success) along with the extracted image. Conversely, when the frame reading process encountered an error, the model returned a null value. If the image is captured accurately, our initial step involves resizing the image. Then let's include it in the list. Now we create the actual data derived from the sequence pictures. For creating data, we create the list of videos for the current class; once we have the list of all the videos, we pass the videos one by one for extracting the frames from the given video, and once the list of frames is obtained, we verify whether the count of frames matches the count of images.

In Figure 4.5, we illustrate our ConvLSTM-based model for video classification. This architecture combines CNNs and LSTMs to effectively and efficiently recapture both spatial and temporal information contained in video data. Within this design, convolutional layers retrieve spatial features from single frames, whereas LSTM layers seamlessly manipulate these characteristics to effectively learn about the temporal dynamical patterns throughout the video frames. With this integrated approach, our model can reach a higher level of classification accuracy by gaining an accurate understanding over time of the contexts and relationships between the images. Together, the design described in the figure outlines various elements of the model, notably the input preprocessing, the ConvLSTM layer stacks. and the ultimate output for class matching, which together contribute to robust performance in video classification tasks.

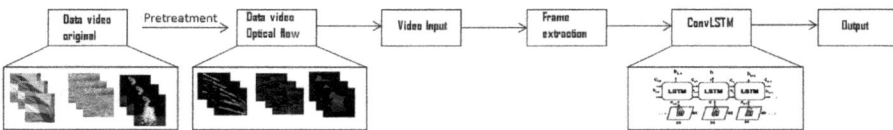

FIGURE 4.5   ConvLSTM-based video classification.

It is important to note that the training of such models can be time-consuming, especially considering the number of categories and the volume of videos within each category. In our situation, using a personal laptop, training our model required an average of 45 hours.

4.5 RESULTS AND DISCUSSION

To evaluate the performance of our model, we used standard metrics such as precision, recall, F1-measure, and accuracy.

4.5.1 EVALUATION METRICS

The efficacy of the proposed method is evaluated using several performance metrics, including precision, recall, F1-measure, and accuracy. These metrics are computed based on the counts of actual positive (AP), false-positive (FP), false-negative (FN), and actual negative (AN) instances. Accuracy is measured by the number of instances that are correctly assigned to the normal and attack classes. Accuracy is attained by averaging the occurrences that were correctly identified and dividing by the total instances as shown in Equation (4.10):

$$Accuracy = \frac{TP + TN}{TP + FP + FN + TN} \qquad (4.10)$$

Evaluation of true positive (TP) and false-positive (FP) entities is the goal of Precision as shown in Equation (4.11) [21, 22]:

$$Precision = \frac{TP}{TP + FP} \qquad (4.11)$$

Recall assessing the ability of a model to correctly identify true positive (TP) instances in comparison to the total number of false-negative (FN) instances that remain uncategorized. Equation (4.12) illustrates the mathematical representation of recollection [22]:

$$Recall = \frac{TP}{TP + FN} \qquad (4.12)$$

Performance evaluations might not always be acceptable for accuracy and recall. For instance, if a mining algorithm had great precision but low recall, a different approach would be required. The average recall and precision provided by the F1-score could be used to fix this issue. Equation (4.13) can be used to compute the F1-score [23, 27, 28]:

$$F1\text{-}Score = \frac{2*(Recall*Precision)}{(Recall + Precision)} \qquad (4.13)$$

4.6 RESULTS AND DISCUSSION

In order to assess how the number of frames influences the performance of the classifier and identify the optimal frame count for improved results, our study involved training the model across varying frame counts: specifically, 30, 40, 50, 60, 70, and 80 frames per video. We partitioned the data into 80% for training purposes and reserved the remaining 20% for testing the classifier. Table 4.2 provides a comprehensive overview of the classifier's performance, evaluated in terms of recall, accuracy, F1-score, and precision.

The evaluation results reveal the performance of the classification model across different lengths of frame sequences. Starting at a range of 30 frames, both the accuracy and F1-score are 0.50, indicating a moderate level of classification accuracy. As the frame length increases to 40 frames, there is a significant improvement in both accuracy and F1-score, reaching 0.59, indicating enhanced model effectiveness. However, with further increase to 50, 60, and 70 frames, the accuracy drops to 0.43, while the F1-score drops to 0.42, indicating a deterioration in classification performance. Interestingly, using 40 frames leads to higher performance for our model. These results emphasize the importance of optimizing frame length in video classification tasks to achieve optimal performance.

4.7 CONCLUSION AND PERSPECTIVES

In conclusion, the evaluation of classification performance across different frame lengths provides valuable insights into the effectiveness of the model in analyzing video data. While shorter frame sequences exhibit moderate accuracy and F1-scores, suggesting a baseline level of classification capability, there is a notable improvement in performance as the frame length extends to 40 frames. However, beyond this threshold, the performance metrics display a trend of fluctuation, with diminishing returns observed in longer frame sequences. These findings emphasize the importance of carefully selecting the frame length in video classification tasks to balance computational efficiency with classification accuracy. Moreover, they highlight the

TABLE 4.2
Evaluation Results of the ConvLSTM-Based Model Using the Dataset

	Accuracy	Precision	Recall	F1-Score
30 Frames	0.50	0.52	0.50	0.50
40 Frames	0.59	0.61	0.59	0.59
50 Frames	0.51	0.53	0.51	0.49
60 Frames	0.50	0.53	0.50	0.48
70 Frames	0.45	0.46	0.45	0.43
80 Frames	0.43	0.45	0.43	0.42

need for further exploration and optimization of model parameters to enhance classification performance across a broad spectrum of video durations. Overall, this study contributes to the ongoing research efforts aimed at improving the efficacy of video classification algorithms and their applicability in real-world scenarios.

REFERENCES

[1] M. Benzyane, I. Zeroual, M. Azrour, and S. Agoujil, "Convolutional Long Short-Term Memory Network Model for Dynamic Texture Classification: A Case Study", in *International Conference on Advanced Intelligent Systems for Sustainable Development*, J. Kacprzyk, M. Ezziyyani, and V. E. Balas, Éds., Cham: Springer Nature Switzerland, 2023, pp. 383–395. https://doi.org/10.1007/978-3-031-26384-2_33.

[2] M. Benzyane, M. Azrour, I. Zeroual, and S. Agoujil, "State-of-the-Art Methods for Dynamic Texture Classification: A Comprehensive Review", in *Sustainable and Green Technologies for Water and Environmental Management*, M. Azrour, J. Mabrouki, and A. Guezzaz, Éds., Cham: Springer Nature Switzerland, 2024, pp. 1–13. https://doi.org/10.1007/978-3-031-52419-6_1.

[3] V. Kellokumpu, G. Zhao, and M. Pietikäinen, "Recognition of Human Actions Using Texture Descriptors", *Machine Vision and Applications*, vol. 22, no. 5, pp. 767–780, 2011. https://doi.org/10.1007/s00138-009-0233-8.

[4] Y. Qiao and L. Weng, "Hidden Markov Model Based Dynamic Texture Classification", *IEEE Signal Processing Letters*, vol. 22, no. 4, pp. 509–512, 2015. https://doi.org/10.1109/LSP.2014.2362613.

[5] A. Schodl, R. Szeliski, D. H. Salesin, and I. Essa, "Video Textures", in *Seminal Graphics Papers: Pushing the Boundaries, Volume 2*, 1st ed., vol. 2, New York: Association for Computing Machinery, 2023, pP. 557–570. Consulté le: 23 novembre 2024. [En ligne]. https://doi.org/10.1145/3596711.3596769.

[6] C. Hazman, A. Guezzaz, S. Benkirane, and M. Azrour, "Enhanced IDS with Deep Learning for IoT-Based Smart Cities Security", *Tsinghua Science and Technology*, vol. 29, no. 4, pp. 929–947, 2024, https://doi.org/10.26599/TST.2023.9010033.

[7] H. Hissou, S. Benkirane, A. Guezzaz, M. Azrour, and A. Beni-Hssane, "A Novel Machine Learning Approach for Solar Radiation Estimation", *Sustainability*, vol. 15, no. 13, Art. no. 13, 2023, https://doi.org/10.3390/su151310609.

[8] M. Benzyane, M. Azrour, I. Zeroual, and S. Agoujil, "Exploring the Impact of Convolutions on LSTM Networks for Video Classification", in *Artificial Intelligence, Data Science and Applications*, Y. Farhaoui, A. Hussain, T. Saba, H. Taherdoost, and A. Verma, Éds., Cham: Springer Nature Switzerland, 2024, pp. 21–26, https://doi.org/10.1007/978-3-031-48573-2_4.

[9] G. Doretto, A. Chiuso, Y. N. Wu, and S. Soatto, "Dynamic Textures", *International Journal of Computer Vision*, vol. 51, no. 2, pp. 91–109, 2003, https://doi.org/10.1023/A:1021669406132.

[10] W. N. Gonçalves and O. M. Bruno, "Dynamic Texture Analysis and Segmentation Using Deterministic Partially Self-Avoiding Walks", *Expert Systems with Applications*, vol. 40, no. 11, pp. 4283–4300, 2013, https://doi.org/10.1016/j.eswa.2012.12.092.

[11] K. Bella, et al., "An Efficient Intrusion Detection System for IoT Security Using CNN Decision Forest", *PeerJ Computer Science*, vol. 10, p. e2290, 2024, https://doi.org/10.7717/peerj-cs.2290.

[12] S. Khan, et al., "Manufacturing Industry Based on Dynamic Soft Sensors in Integrated with Feature Representation and Classification Using Fuzzy Logic and Deep Learning Architecture", *International Journal of Advanced Manufacturing Technology*, 2023, https://doi.org/10.1007/s00170-023-11602-y.

[13] G. Zhao and M. Pietikainen, "Dynamic Texture Recognition Using Local Binary Patterns with an Application to Facial Expressions", *IEEE Transactions on Pattern Analysis and Machine Intelligence*, vol. 29, no. 6, pp. 915–928, 2007, https://doi.org/10.1109/TPAMI.2007.1110.

[14] S. Khan, et al., "Enhanced Spatial Stream of Two-Stream Network Using Optical Flow for Human Action Recognition", *Applied Sciences*, vol. 13, no. 14, Art. no. 14, 2023, https://doi.org/10.3390/app13148003.

[15] T. Gao, et al., "Sports Video Classification Method Based on Improved Deep Learning", *Applied Sciences*, vol. 14, no. 2, Art. no. 2, 2024, https://doi.org/10.3390/app14020948.

[16] M. Benzyane, M. Azrour, I. Zeroual, and S. Agoujil, "Investigating the Influence of Convolutional Operations on LSTM Networks in Video Classification", *Data and Metadata*, vol. 2, p. 152, 2023, https://doi.org/10.56294/dm2023152.

[17] D. Sun, S. Roth, J. P. Lewis, and M. J. Black, "Learning Optical Flow", in *Computer Vision – ECCV 2008*, D. Forsyth, P. Torr, and A. Zisserman, Éds., Berlin, Heidelberg: Springer, 2008, pp. 83–97, https://doi.org/10.1007/978-3-540-88690-7_7.

[18] Z. Chen, Y. Quan, R. Xu, L. Jin, and Y. Xu, "Enhancing Texture Representation with Deep Tracing Pattern Encoding", *Pattern Recognition*, vol. 146, p. 109959, 2024, https://doi.org/10.1016/j.patcog.2023.109959.

[19] X. Zeng, X. Zhao, X. Zhong, and G. Liu, "A Survey of Micro-expression Recognition Methods Based on LBP, Optical Flow and Deep Learning", *Neural Process Letters*, vol. 55, no. 5, pp. 5995–6026, 2023, https://doi.org/10.1007/s11063-022-11123-x.

[20] L. Luan, Y. Liu, and H. Sun, "Extracting High-Precision Full-Field Displacement from Videos Via Pixel Matching and Optical Flow", *Journal of Sound and Vibration*, vol. 565, p. 117904, 2023, https://doi.org/10.1016/j.jsv.2023.117904.

[21] Md. A. Uddin, et al., "Deep Learning-Based Human Activity Recognition Using CNN, ConvLSTM, and LRCN", *International Journal of Cognitive Computing in Engineering*, vol. 5, pp. 259–268, 2024, https://doi.org/10.1016/j.ijcce.2024.06.004.

[22] R. Vrskova, P. Kamencay, R. Hudec, and P. Sykora, "A New Deep-Learning Method for Human Activity Recognition", *Sensors*, vol. 23, no. 5, Art. no. 5, 2023, https://doi.org/10.3390/s23052816.

[23] R. Chaganti, A. Mourade, V. Ravi, N. Vemprala, A. Dua, and B. Bhushan, "A Particle Swarm Optimization and Deep Learning Approach for Intrusion Detection System in Internet of Medical Things", *Sustainability*, vol. 14, no. 19, p. 12828, 2022, https://doi.org/10.3390/su141912828.

[24] M. K. Boutahir, Y. Farhaoui, M. Azrour, I. Zeroual, and A. El Allaoui, "Effect of Feature Selection on the Prediction of Direct Normal Irradiance", *Big Data Mining and Analytics*, vol. 5, no. 4, pp. 309–317, 2022, https://doi.org/10.26599/BDMA.2022.9020003.

[25] H. Attou, et al., "Towards an Intelligent Intrusion Detection System to Detect Malicious Activities in Cloud Computing", *Applied Sciences*, vol. 13, no. 17, p. 9588, 2023.

[26] A. Toktarova, et al., "Hate Speech Detection in Social Networks Using Machine Learning and Deep Learning Methods", *International Journal of Advanced Computer Science and Applications*, vol. 14, no. 5, pp. 396–406, 2023.

[27] S. Dargaoui, M. Azrour, A. Allaoui, A. Guezzaz, A. Alabdulatif, and A. Alnajim, "Internet of Things Authentication Protocols: Comparative Study", *CMC*, vol. 79, no. 1, pp. 65–91, 2024, https://doi.org/10.32604/cmc.2024.047625.

[28] S. Dargaoui, M. Azrour, A. El Allaoui, A. Guezzaz, A. Alabdulatif, and A. Alnajim, "An Exhaustive Survey on Authentication Classes in the IoT Environments", *Indonesian Journal of Electrical Engineering and Informatics (IJEEI)*, vol. 12, no. 1, Art. no. 1, 2024, https://doi.org/10.52549/ijeei.v12i1.5170.

5 Digital Guardians
Revolutionizing Pest Management with AI and IoT for a Greener Future

Sreedeep Dey and Subhasis Roy

5.1 INTRODUCTION

Insect pests cause losses in agriculture locally as well as globally. To limit crop damage caused by these pests, we need to control them. We have to detect and classify these pests properly [1]. With that accuracy, it becomes possible to identify distinct species. This identification aids us in establishing pest control measures that are more appropriate. As the traditional pest monitoring systems were dependent upon human skill sets for the process of identification, in most of the cases they have been proven inaccurate. These processes consumed too much time and were also a bit expensive [2–5]. The on-growing interest in automated pest recognition has been arising from that need. They are constant and charge a lower cost to monitor. On the other hand, we know that abusive pesticides also harm the organisms that are beneficial. This hampers the ecosystems. It emphasizes why we have to focus more on pest management measures.

When we combine AI with computer vision and Internet of Things (IoT) [6, 7], it becomes a viable approach that helps the community with automated pest identification. With those systems, we can classify those pests with the use of cameras and sensors. This technology allows us to increase the efficiency of pest monitoring, eliminating the need for human identification and providing real-time analysis. AI is a potent candidate that transforms pest management. The solutions AI provide are more precise, persist in the long term, and are environmentally friendly. Smart pest monitoring (SPM) is a subfield of Integrated Pest Management (IPM) [8]. SPM automates pest data gathering by leveraging AI, IoT, and big data. This allows the improvement of early warning systems. As a consequence, pest management decision-making, for example, pesticide application, becomes more improved.

With promising AI technologies like ML [9, 10], computer vision, and data analytics, we can study large amounts of datasets. These insights provide accurate visions and pest management tactics benefit. Utilizing AI, farmers can monitor pest numbers and density in their fields, enabling them to make informed pest control decisions.

DOI: 10.1201/9781003527664-5

With those decisions, they can reduce their crop losses by optimizing the use of pesticides. AI is capable of delivering insights that are, with fine details, data-driven. This reduces the necessity of chemical treatments. As a result, the environment faces less damage and it helps to conserve beneficial animals. These beneficial predators and parasitoids can then leverage the management of pest populations naturally.

Among several benefits, there are still some hurdles to face in the case of pest management using AI. To overcome those issues, we have to focus on data privacy and related legal compliances, and obtain cross-disciplinary collaboration between agronomy, ecology, and computer science. We have to make sure that small-scale farmers and farmers in resource-constrained locations like India, are also aware of the benefits of AI technology and can access and use them.

For example, in cotton farming, farmers have encountered many difficulties when they relied upon traditional pest scouting methods. They were dependent upon professionals who located and identified harmful bugs manually. This technique was time-consuming and error-prone. Largely populated pest species in cotton ecosystems, which are approximately 1,000 around the world and up to 125 per country, fuel these error rates [11]. There are several species of pests that are almost impossible to identify with the naked human eye. Identifying them without using AI and IoT-based sensor technologies can result in considerably higher use of pesticides. They can give rise to resistance properties, pose serious health hazards, and potentially damage the environment [12, 13]. Not only that, when scouting manually the application of selective pesticides may be overlooked [14, 15]. This is especially important for the preservation of beneficial organisms that help to control pests in a natural way.

There are several AI technologies. These include K-nearest neighbors (KNN), logistic regression, decision trees, support vector machines (SVMs), and deep convolutional neural networks (CNNs) [16]. These technologies assist farmers in properly detecting the pests and classifying them in appropriate groups [17]. These models can be used to recognize such species that are either relatively smaller and/or pose difficulty in being detected [18]. AI is capable of evaluating photos obtained by cameras, which are not very expensive. These low-cost cameras make AI more utilizable in pest management [19].

In this chapter, we explain why incorporating AI into pest management seems to be an important step to improving food security across the world and ultimately reducing the environmental effects that can be posed through agriculture. Figure 5.1 highlights the AI tools that are used to manage pests and disease. Figure 5.1A shows the AI classification algorithms. As can be seen, it is dominated by self-organizing maps and wavelet-based methods. Figure 5.1B illustrates image segmentation algorithms. With it, clustering methods are taking precedence. Figure 5.1C shows the feature extraction techniques. It includes PCA and gradient-based features. Sensing technologies are summarized in Figure 5.1D. Imaging and heat sensors are the most dominant sensors. Figure 5.1E shows insect pests like whiteflies and bollworms. Figure 5.1F lists common cotton diseases and include root rot and bacterial infections.

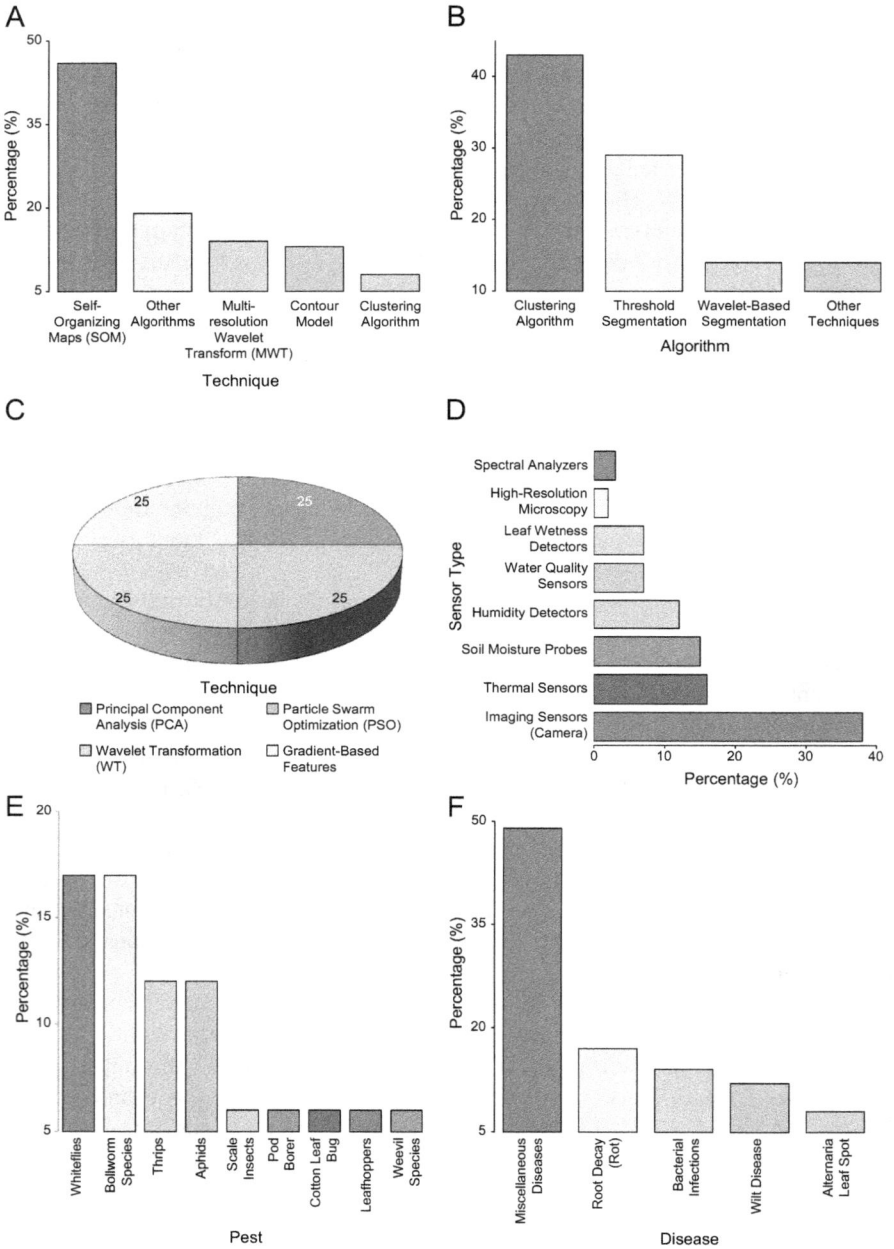

FIGURE 5.1   (A) AI-based classification techniques for pest and disease identification. (B) Image segmentation algorithms applied in pest detection. (C) Key feature-extraction methods used in pest identification. (D) Sensing technologies for pest and disease monitoring. (E) Insect pests of cotton crops. (F) Cotton diseases under investigation.

5.2 INNOVATIVE BIOLOGICAL AND PHEROMONAL PEST CONTROL METHODS

5.2.1 Biological Pest Control: A Sustainable Approach

In an attempt to control insects, weeds, and plant diseases, a promising approach is biological pest control. This system employs naturally existing predators, parasites, and even pathogens. It is an environmentally beneficial method. Synthetic pesticides can harm non-target animals. They misbalance the ecosystems. Biological pest control reduces their need. Among biological management approaches, is introducing helpful insects, such as ladybugs that attack aphids. Microbe-based sprays are also being used. For example, *Bacillus thuringiensis* (Bt), which is a bacterium that becomes harmful to certain types of pests when they engulf and digest them [20].

5.2.2 Key Benefits of Biological Pest Control

(1) *Environmental safety:* As there is no use of synthetic pesticides, it causes less harm to the beneficial species, maintaining a balanced environment.
(2) *Sustainability:* Populations of predators and prey stay equilibrated, promoting IPM.
(3) *Reduce resistance property:* Some of the common pests, like green peach aphids, are becoming resistant to pesticides. From that point, biological control is an alternative. It targets resistant pests, for example, the Colorado potato beetle which has shown resistance against 70 synthetic insecticides [21].

Biological pest control presents several benefits. But practically, in comparison to chemical treatments, for this kind of control system more meticulous planning is needed. The monitoring system is also a little bit complicated.

5.2.3 Pheromonal Pest Control: Precision Targeting

Insect hormones are used in pheromonal pest control. These hormones have the ability to either disrupt pests and/or confound them. Populations of insects usually use these pheromones for communicating with each other as a natural instinct. This special type of pest control uses these pheromones to lure and trap them. Sometimes this control also disturbs their mating habits. For example, moths or cockroaches can be attracted by pheromone-based traps. But mating disruption tactics work differently. They divert male insects away from female partners which results in lowering pest populations.

5.2.4 Advantages of Pheromonal Pest Control

(1) *Targeted action:* Pheromones are also specific to the species. Similarly, the impact on non-target insects is reduced and wildlife is protected.
(2) *Versatility:* One of the most notable advantages of pheromonal traps and sprays is that they are versatile. They are effective at the same frequency across all phases of the lifetime. They can be used in any kind of flexible setting as well as in our homes and farms.

(3) *Low environmental impact:* As synthetic chemicals are not being used, the environment al impact is reduced with pheromonal methods [22].

5.2.5   RECEPTOR INTERFERENCE: A NEW FRONTIER IN PEST CONTROL

Hormones and pheromones work as some of the most important signals for insects. These signals affect their most crucial vital behaviors, which are feeding and mating. Receptor interference prevents the insects from detecting those signals that play an important role in their lives. This emerging technology provides a tailored and environment-friendly alternative to relying only on petrochemical pesticides [23].

5.2.6   STERILE INSECT TECHNIQUE (SIT): REPRODUCTIVE CONTROL

The SIT releases male insects into the wild. These male insects are sterilized males. There they mate with wild female partners. As a result, they lay sterile eggs. SIT technology mainly focuses on chemical-free pest control. This approach targets reducing populations over not only one but several seasons. There are many successful examples included. Among them, the management of the Mediterranean fruit fly and the screwworm have been most impactful [24, 25]. Both of them have caused major damage in agriculture nationally and internationally.

As far as we can understand, this novel approach is effective as a solution that can provide minimum impact on the environment. However, as our target is to yield long-term benefits, they need to be used very consistently.

5.3   EXPLORING AI CLASSIFICATION: A FUNCTIONAL AND LEARNING-BASED PERSPECTIVE

AI can be classified according to (1) its capabilities, (2) the learning methods that are being used, and (3) the respective domains where it is being applied. Usually we can classify AI as either narrow AI (also known as weak AI) or general AI (also known as strong AI). Narrow AI is specific to the given task, which can excel in a few areas. These include (1) image recognition, (2) virtual assistants, and (3) system recommendations [26]. Unfortunately, there is a lack in AI's broad understanding. On the other hand, general AI works like us – using human intelligence. It becomes adaptable to new tasks and then learns across domains [27].

AI can be divided into three learning approaches: (1) supervised, (2) unsupervised, and (3) reinforcement learning. (1) *Supervised learning* maps inputs to outputs, it trains itself on labeled datasets. (2) *Unsupervised learning* mainly identifies patterns that are in unlabeled data. (3) In *reinforcement learning*, agents learn optimal tactics through interactions with their environment, seeking exchange for either rewards or penalties [28]. Kar et al. used self-supervised learning (SSL) techniques, classifying insect pests with the help of ResNet models [29].

AI can be applied in specialized domains, such as in Natural Language Processing (NLP) that can comprehend human language; it can be used in computer vision that can analyze images and video. Robots can be used for autonomous task performance [30]. With NLP, farmers can diagnose diseases in agricultural fields. With computer vision, they can do facial identification and can analyze medical pictures [31].

Lastly, ML and deep learning (DL) are two subsets of AI that are focused on techniques that make systems compatible to learn from data. DL takes advantage of multilayered neural networks [32]. These networks are fashioned like the human brain; and with minimal intervention, they develop autonomous judgments.

5.3.1 REVOLUTIONIZING PEST CONTROL WITH ML

ML helps the systems learn from data. To do that, they do not demand any explicit programming. This feature makes it an effective tool that can identify and monitor pests in the field. A diverse range of data, including photos and readings from sensors, can be evaluated [10, 33, 34]. In that way, we can fluently detect pest infestations and manage them with great efficiency [35].

5.3.1.1 Image Recognition and Classification

Photos of crops can be recognized and classified by ML systems. These models can detect the pests and damage they cause. Drones are flying devices that are equipped with high-resolution cameras. These drones take photographs of fields from different positions. These photos are analyzed to identify anomalies. Penn State University has used "Plant Village" software. It uses ML models. From photos uploaded by the user or farmer, it identifies pests and diseases [36]. Similarly, by using ML, Wang and Xie categorized 24 insect species.

5.3.1.2 Smart Sensors and IoT Integration

Real-time data can be obtained from different types of sensor networks and IoT devices. These devices and those obtained data allow us to monitor pests.

(1) *Low-power cameras* first collect images of the pest that has been trapped. Then the images are transmitted to be analyzed. After that, the farmers locate the place of infestations.
(2) *Thermal sensors* measure plant and soil spectral signatures to identify variations in them. In this way, they detect pest infestations.
(3) Chlorophyll patterns are the main focus of *fluorescence sensors* for detecting infections and causative pests.
(4) Different insects and rodents develop different sound waves. *Acoustic sensors* analyze those formed waves and detect pests.
(5) *Gas sensors* analyze volatile chemicals, which can detect stress in plants induced by pests.

There are some really promising systems, like the "Spornado" bug monitoring system and the "Trapview" system [37, 38]. These systems use IoT sensors and ML. They monitor pest populations in real time, which help the farmers to be aware of relevant insights.

5.3.1.3 Automated Pest Identification

How can we make the pest identification automated? The answer is the use of ML algorithms. These algorithms analyze inputs. These inputs include pictures, DNA, or acoustic data. There are readily available tools like "PestID" software that rely on smartphone cameras to identify pests swiftly and expedite the whole complex procedure [39]. Spiesman et al. used ML and successfully identified 36 bumblebee species [39]. The Angel invariant Gabor algorithm recognized Lepidopteran insects; notably, the accuracy was 75% [38].

5.3.1.4 Predictive Pest Modeling and Forecasting

Studying previously available data on weather, climate, water, rain, floods, agriculture, and pest infestations helps ML forecast breakouts of the pest in an early stage [40]. With this predictive modeling, farmers can anticipate infestations. Acquiring possible precautions becomes easier for them. CSIRO has their "Pest Forecast" technology. Using ML it helps to predict the activities of the pest. Farmers can reduce crop loss and optimize the use of pesticides in the field with CSIRO's promising technology. These solutions powered by ML are working like a game changer in pest management. They are boosting the detection accuracy, enabling farmers to equip actions promptly.

5.3.2 Unleashing Deep Learning (DL) for Pest Detection

DL employs multilayered neural networks. Particularly in pest monitoring in agricultural fields, this DL has tremendously transformed the process of picture recognition. DL models process raw data. With those processed data, they excel at the process of pest and plant disease detection very accurately.

5.3.2.1 Image Recognition with DL

A particular DL model called CNN is very efficient for analyzing visual data. They use photos, which can be taken by drones, smartphones, or cameras mounted in the field for insect infestation detection purposes. CNNs are developed for image identification and classification tasks. Pest species can be distinguished and plant health issues can be detected. The YOLO v5 algorithm identifies insects of crops and counts their numbers. Under greenhouse settings, they can detect pests like aphids, fruit flies, and thrips with a higher accuracy of 96% [41]. The "DeepAgro" platform works similarly with DL to detect insect infestations and diseases in crops [42]. A flowchart with decision branches is illustrated in Figure 5.2. It helps for model selection and pest identification. Sequential flow of activities is also shown based on the extended structure.

5.3.2.2 Sensor Fusion for Pest Monitoring

DL combines data from various types of sensors. These sensors include temperature, humidity, and motion sensors. With them, farmers understand pest activity comprehensively. The data can be obtained from weather stations, soil moisture sensors, and insect traps. Then, by combining them prior to using DL models, pest infestations can be anticipated. For example, in vineyards, camera sensors are used with DL-based models. This allows users to identify *Bactrocera oleae* and *Ceratitis capitata* pests with 93% accurately [43].

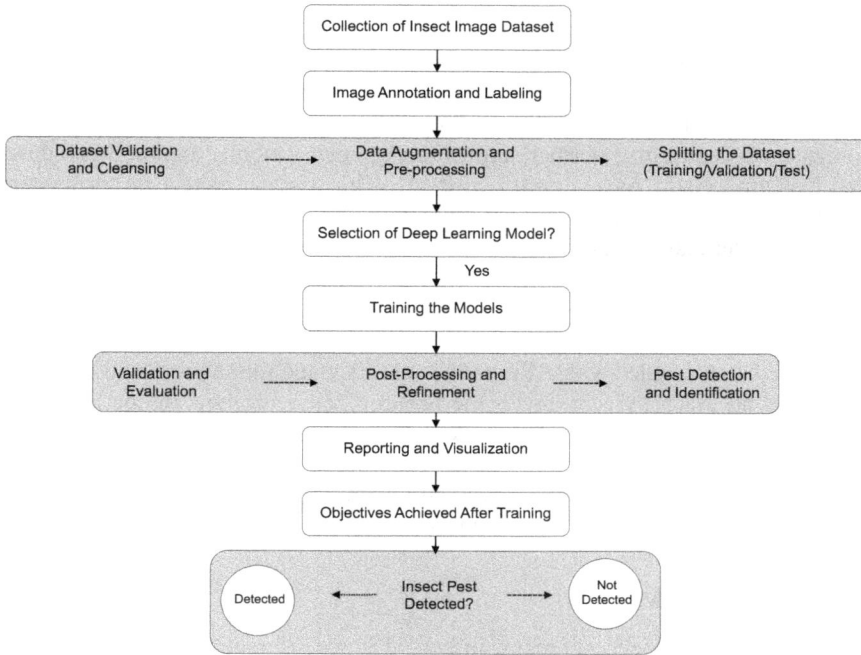

FIGURE 5.2 Flowchart illustrating the process of insect pest detection and identification using DL models, from dataset collection to outcome evaluation.

5.3.2.3 Automated Pest Detection with IoT

In agricultural fields, IoT devices that are combined with DL algorithms deliver insect monitoring in real time. These systems use cameras, microphones, and environmental sensors. With those devices, they maintain continuity to monitor the activity of the insect. Pest populations can be monitored with the "Trapview" system. Real-time notifications are sent out with them. Advanced acoustic sensors use DL algorithms with sound detection. In this case, they have achieved over 99% accuracy, where they recognized all together 800 pest species. They outperformed all the relevant standard models, including the highly hyped ResNet-50 [44].

As DLs are able to learn from complex data and discern patterns, they are becoming a promising tool that can be applied in the future for automated pest identification and also help to control the process.

5.4 AI REVOLUTION IN PEST DETECTION AND MONITORING

Pest management is becoming altered with AI. AI suggests better methods with which we can protect our agricultural output with subsequent detection, monitoring, and control of pests. In this section we provide a short throughput at its main applications.

5.4.1 EARLY DETECTION

AI uses satellite imagery, drones, and sensors to detect pests prior to the appearance of signs. AI systems use ML models. Decision trees, support vector machines, and CNNs are among those models that can detect the existence of anomalies that compromise crop health in an earlier stage. For example, a mobile app has been developed by Tamil Nadu Agricultural University. This app can detect fall armyworm infestations in maize. The detection is more than 93% accurate. Technically, this app is using transfer learning and DL approaches [45].

5.4.2 PRECISION MONITORING

Drones are being coupled with AI-powered sensors to monitor agricultural fields with astonishing accuracy. AI scrutinizes the health of the crop, how the pest migrates, and focuses on environmental data to determine infestation zones. SPM systems in vineyards are rapidly using segmentation techniques. This helps farmers to identify the grape moth better than before. Successful identification helps to reduce the use of pesticides and increase the overall crop yields.

5.4.3 PREDICTIVE MODELS

When AI forecasts insect outbreaks, historical as well as real-time data (like trends of weather and pest life cycles) is used. With this forecast, farmers are able to utilize proactive pest management measures. We have discussed in the previous section that IoT cameras and sensors rapidly identify pests in the field. This identification enhances effective Integrated Pest Management (IPM), therefore reducing pesticide use and lessening the environmental impact.

5.4.4 AUTOMATION AND ROBOTICS

Robots and autonomous devices powered by AI automate pest monitoring and control. As a result, the requirements of labor are decreased and subsequent production is increased. These systems first collect the data. Second, they pinpoint hotspots of the pest. Lastly, targeted actions are enabled. Sticky paper traps are a kind of automated trap. They are usually paired with CNN-based image classifiers. With those traps, crucial information on pest activity can be obtained in real time. Figure 5.3 depicts optimizing edge intelligence systems for real-time pest and weather monitoring. This also promotes sustainable agriculture practices. This occurs by improving data processing, decision support, and the improvement of ongoing models.

5.4.5 DATA-DRIVEN DECISION-MAKING

Nowadays, farmers are optimizing pest management as they take advantage of AI-powered drones and decision-making tools. These specialized drones use AI and edge intelligence systems. They identify pests in real time and recommend the routes by which pesticides can be sprayed.

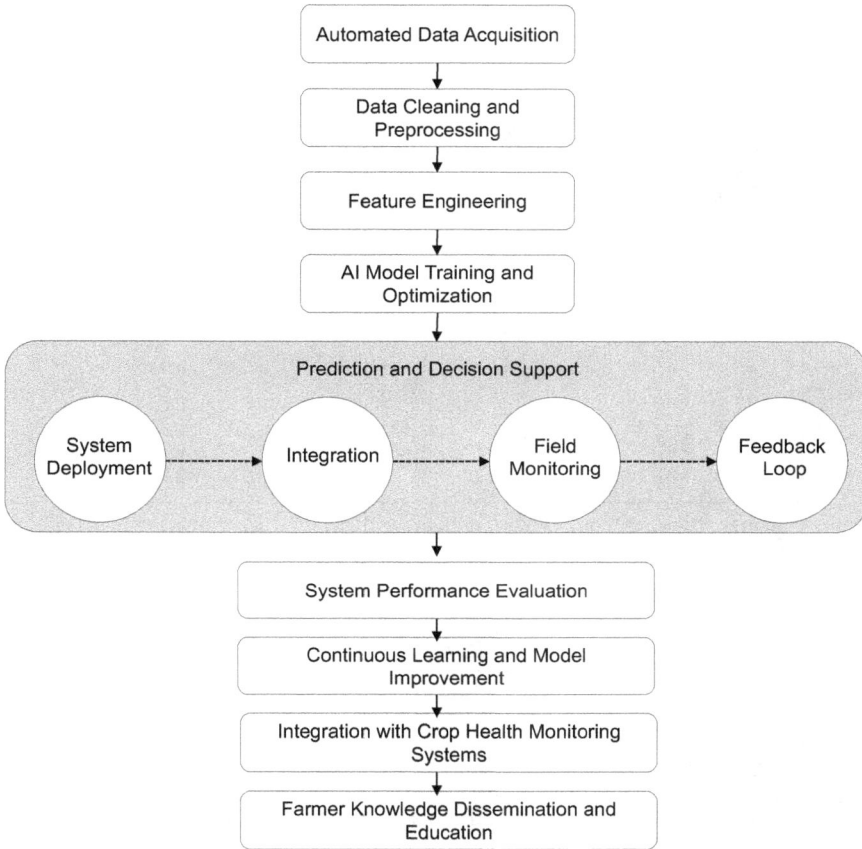

FIGURE 5.3   Stages involved in smart pest and weather recognition for sustainable farming practices.

5.5   AI-ENHANCED SENSORS IN PEST DETECTION

Banerjee et al. have described how the prospective of pest management has changed with AI-integrated sensors [46]. These sensors translate biological and environmental interactions into actionable data. These sensors are like the "eyes and ears" of an AI system. They collect data with numerous technologies and devices including cameras, microphones, LiDAR, and inertial measurement units [47]. For finding trends and filtering the noise out, AI algorithms are applied to the obtained data. This is done frequently in real time. ML also assists AI systems while interpreting this data. After that, AI becomes able to make intelligent decisions. When sensors are integrated with AI, ML, and DL models, they are able to provide some really effective solutions that detect diseases and their causative pests in agriculture. Table 5.1 provides a comprehensive overview where AI and sensor technologies integrate together for controlling pests. It also highlights

TABLE 5.1

Integrated Overview of AI and Sensor Solutions in Pest Control

Application/Sensor Type	Crop Type	Target Pest/ Disease	AI Technique/ Sensor Technology	Detection Precision/ Method	Advantages	Limitations
Aphid detection in wheat cultivation	Wheat	Aphids	Support vector machine (SVM), maximally stable extremal regions (MSERs), histogram of oriented gradients (HOGs)	High precision with detailed feature analysis	Early identification reduces crop damage and boosts yield	Needs high-quality images for precise detection
General pest identification	All crops	Various insect pests	Deep convolutional neural network (DCNN)	High precision in recognizing multiple pest types	Efficient monitoring across a variety of crops	Limited training data for specific pests
Counting and recognizing flying insects	All crops	Flying insects	Raspberry Pi with image processing	Real-time counting and identification	Cost-effective and portable monitoring solution	Performance may be affected by environmental conditions
PEN3 electronic nose	Various crops	Brown rice plant hoppers	PCA, back-propagation neural network (BPNN), probabilistic neural network (PNN)	Age: 96.67% Count: 64.67%	Highly precise insect detection	Limited applicability to specific pest types
MOS sensor array for disease detection	Various crops	*Pectobacterium carotovorum*	PCA	80–100% precision	Quick disease detection enables timely action	May require calibration for various crops
Sweeping electronic nose system	Various crops	*Bactrocera dorsalis*	PCA and LCA	100% precision	Comprehensive pest identification	Complexity in setup and data analysis
Alpha MOS Fox 3000	Various crops	*Citrus tristeza* virus	Support vector classifier (SVC)	97.67% precision	Effective for identifying specific diseases	Cost may be a barrier for small-scale farmers

various types of applications, crops of target, and pests including advantages and limitations of each methodology.

5.5.1 AI-Powered Pest Detection with Sensors

Researchers have employed high-resolution cameras and DL algorithms while creating cutting-edge pest detection systems that are based on AI. These technologies perform field scanning autonomously. They detect insect infestations in real time. Farmers have become careful when applying pesticides in the field. This targeting technology is much more precise and reduces overall use of pesticides, therefore lowering substantial harm that may affect the crops. AI-enabled smart traps in orchards are useful for delivering real-time data about how many insects are populating. Farmers with those insights can work on improving pest control techniques. In addition, advanced drones are being armored with multispectral cameras and LiDAR sensors [48]. These kinds of drones are beneficial for large-scale surveillance. The AI detects infestations with proper evaluation of aerial imagery.

5.5.2 Chemical Sensors for Plant Volatiles and Semiochemicals

There are marketed available chemical sensors that can detect volatile organic compounds (VOCs) the pests or plants secrete. These sensors detect pests at an early stage. Chemo-, bio-, and nanosensors are included in these types of sensors [49]. Interactions with VOCs are translated into detectable signals as the working principle of these sensors. Electronic noses are devices that are equipped with some gas sensors. These devices detect pests using VOCs. Researchers have effectively diagnosed pest diseases like *Ralstonia solanacearum* in potatoes with PEN3, an electronic nose [50]. On paddy crops, the detection of the number and age of brown rice plant hoppers was more than 96% accurate [51].

Furthermore, when researchers applied principal component analysis (PCA) and linear discriminant analysis (LCA), they observed that the sweeping electronic nose system is able to identify *Bactroicera dorsalis* in citrus fruits [52]. The identification is while 100% accuracy in identification is the goal, it's important to understand that it's often not achievable in real-world scenarios accurate. As these sensor-based AI technologies can detect at an early stage and target with precision, they seem to be a support system that may appear as an ecologically responsible pest management phenomenon in the field of future agriculture. Figure 5.4 depicts several applications of AI and sensor technology that help in agricultural and pest management. Figure 5.4A depicts the division of agricultural systems and apps. Here, farm management and weather forecasting are taking the lead. Figure 5.2B describes the sensors used in smart systems. Here it can be seen that temperature and humidity sensors are the most utilized ones. Figure 5.4C projects the expansion of AI-based pest management apps, which expects a significant increase in market value for insect, rodent, wildlife, and termite management over the next decade.

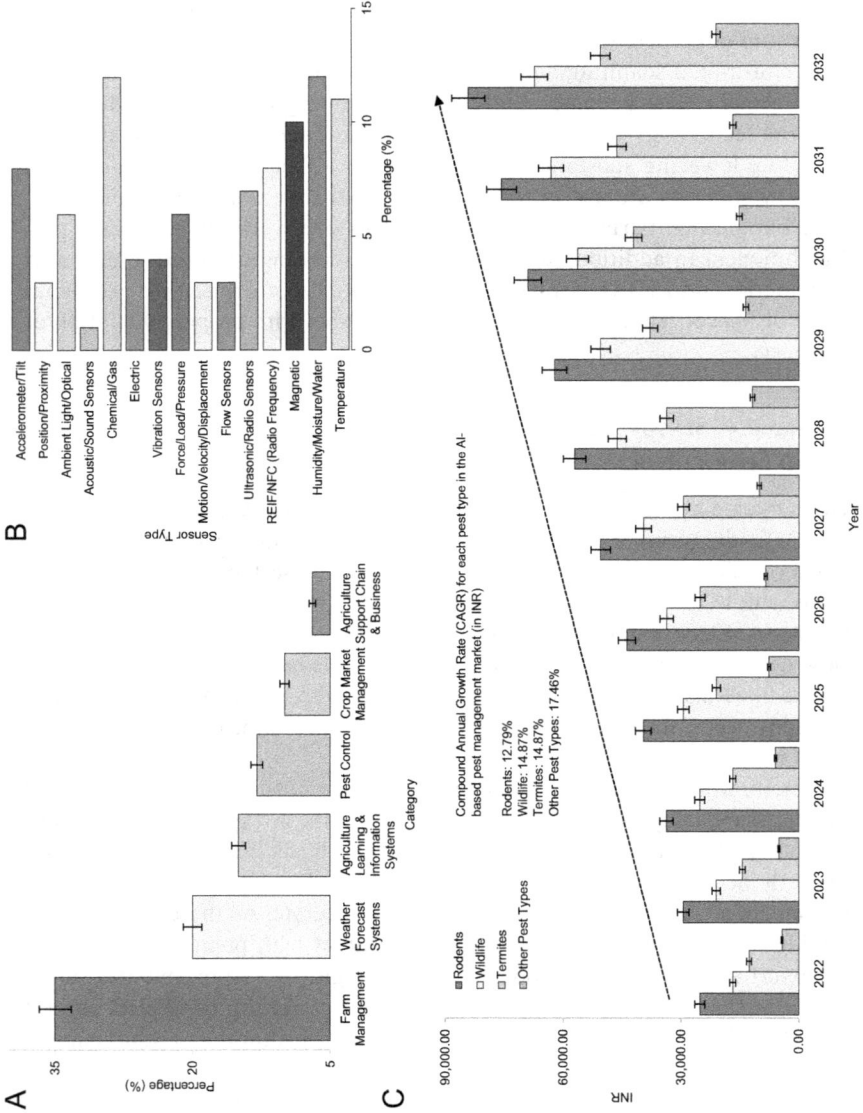

FIGURE 5.4 (A) Distribution of various agricultural system applications. (B) Smart technology sensor applications. (C) Projected growth in the global AI-based pest management app market by pest type from 2022 to 2032.

5.6 KEY MOBILE APPS FOR PEST MANAGEMENT IN AGRICULTURE

(1) *Plant Village:* Plant Village was developed by Penn State University. It is completely a smartphone application. This app provides solutions for pest and disease management for various types of crops. Its database of crop-related issues is sufficiently large to recommend control strategies that are specific to the crops.

(2) *PestID:* The University of Georgia is the developer of the PestID app. Using AI, this app identifies insect pests by detecting photographs taken with a smartphone. The process is very rapid. PestID provides important information about pest biology and acquires the necessary treatment options.

(3) *PestNet:* The FAO itself created PestNet. Farmers here contribute pest and disease photographs. Then these photographs are verified by experts and pest control measures are recommended.

(4) *IPM Toolkit:* It includes (i) guidelines for pest identification, (ii) tools for monitoring, and (iii) recommendations for control strategies – these are the three inclusions in UC IPM's IPM Toolkit that promote sustainable practices.

(5) *FieldScout GreenIndex+:* This software was created by Spectrum Technologies. In this software, remote sensing checks the health of the crop and diagnoses any appearance of stress and damage by insects, which is done through photographs taken by either a smartphone or drone.

(6) *ScoutPro:* ScoutPro is an application for pest and crop health monitoring. Farmers can record their field observations with this app. And in return the app provides them with pest management advice.

(7) *Agrian Mobile:* Agrian Mobile is an impactful app designed for professionals. It not only offerd pest management and crop scouting but also tracked real-time spray, and is able to alert farmers with weather and insect alarms.

(8) *FieldX Scout:* In crops like maize and soybeans, FieldX Scout offers monitoring of insect pressures. With it, users record their day-to-day field reports and receive advice for pest control [38].

5.7 AI SUCCESS STORIES IN AGRICULTURE AND PEST MANAGEMENT

Until now we have provided a basic idea about how using complex algorithms, AI can make substantial advances that ease the process of pest detection and its management. Trimble's Farmer Core software is combining AI and ML for evaluating data directly from farmyards. These data include soil samples, weather patterns, and crop health imaging. With these insights, farmers can improve crop productivity and become more efficient [53]. Taranis is another platform powered by AI. It detects pests and diseases using a combination of weather data and satellite-derived photos. With it, farmers can make data-driven decisions [54].

Alphabet is appearing with their Verily Debug Project. This project aims to analyze environmental parameters and mosquito populations. These data help to predict outbreaks of mosquito-borne diseases like Zika. FarmSense's FlightSensor is an AI and cloud-based application, which can track insects around the clock and classify them accordingly. The accuracy of pest management stays maintained with it. In Africa, the ResNet-50 ML model was employed to identify pest-affected areas. After successful identification, they forecasted outbreaks of the fall armyworm. Future instances were also forecasted [55].

5.8 INNOVATIVE PEST CONTROL SOLUTIONS: SCIENCE-BACKED APPROACHES

Pests are more than a simple nuisance. They spread hazardous disease, taint food sources, and cause damage to resources and properties. Chemicals of traditional pest management methods harmed not only harmful pests but also humans, harmless or beneficial insects, and animals, and ultimately polluted the environment. We are fortunate that the industry is being transformed by improvements in pest control. These improvements are safer and pose as better effective alternatives than the traditional ones.

5.9 IoT IN PEST MANAGEMENT: SMART SOLUTIONS

Physical things are connecting the IoT with sensors that collect data and share it via the Internet. For pest control, IoT enables smart traps and electronic monitoring devices so that they can identify pests and track their activity in the field.

5.9.1 KEY BENEFITS OF IoT PEST CONTROL

(1) *Efficiency:* Automated pest tracking has made the process more efficient. Therefore, technicians can focus on other responsibilities that are more important in agriculture.
(2) *Cost savings:* As the diagnosis is being performed in an early stage, there is a savings in costs and regular inspections and treatments are no longer necessary.
(3) *Environmental sustainability:* As data is being monitored continuously, we can enable tailored chemical applications, reducing the use of pesticides.

IoT systems are usually comprised of two things: (1) A network of sensors, (2) traps for collecting information about environmental conditions and activity of the pest. This data is transmitted to a central hub that detects pests and activates traps. If needed, personnel are notified [56].

5.9.2 EXAMPLES OF IoT SOLUTIONS

(1) *Anticimex SMART:* This is a digital rodent management device. Using predictive algorithms, it detects populations of rats thus avoiding infestations before they occur.
(2) *Rentokil PestConnect:* This system uses infrared sensors. Facilities are monitored around the clock. Real-time alerts are generated. As a result, technicians are able to respond quickly.

5.10 APPLICATIONS OF AI IN PEST CONTROL

5.10.1 AUTOMATED DETECTION

Information about heat, movement, and sound is collected by AI systems using cameras and sensors. ML algorithms utilize this collected data to identify the pests. They give suggestions about probable treatment options. Cotton producers combine AI detection with pheromone traps. This strategy is effective for regulating bollworm infestations.

5.10.2 PREDICTIVE MODELING

AI processes large amounts of data, including weather trends and pest life cycles. Then AI forecasts insect outbreaks. Using this technology and activity data, Anticimex SMART anticipates rodent numbers.

5.10.3 BENEFITS OF PREDICTIVE MODELING

(1) *Proactive pest control:* Technicians can intervene far before the infestations start. Improving reaction time and accuracy.
(2) *Reduced chemical use:* Technicians are provided with the exact required estimates of chemicals, which optimizes the pesticide treatments.
(3) *Cost savings:* Pest control firms can save resources. Farmers can effectively avoid crop losses.

5.11 AI-DRIVEN INNOVATIONS IN PEST CONTROL

5.11.1 SEMIOS AND GOOGLE COLLABORATION: A CASE STUDY

Let's discuss a notable example where AI is practically imprinting its footstep in agriculture. Semios is a precision farming platform that is collaborating with Google. Their collaboration tries to detect obstacles to agriculture and implements possible preventive measures. They have successfully reduced the moth population by 1.5 billion, which increased the yield of almonds [57].

5.11.1.1 About Semios

With Semios, we can leverage a proprietary network of wireless sensors, conduct pest monitoring in real time, and apply biological pest control approaches. These initiatives make agriculture profitable while promoting sustainability. Some of their key offerings include Google Cloud, BigQuery, and TensorFlow.

5.11.1.2 Data-Driven Insights

Semios applies a large network that comprises 500,000 IoT sensors. These IoT sensors are spread across 80,000 acres. They collect data on insects, moisture, and meteorological conditions. Then the data proceeds to be processed in real time. Crop growers and farmers then monitor plant health and determine insect risks more accurately. Google Cloud's BigQuery is being used by Semios. With it, they connect different sources of data. This connection provides complete insights with which pesticide use can be optimized.

5.11.1.3 AI for Pest Detection and Risk Prediction

The advanced ML techniques used by Semios assess microclimate data for identifying pests. Not only that, it has been reported that they are mounting digital cameras into insect traps. These cameras will collect photos that will be processed through AI in later stages. This approach helps the growers in many ways. They can get real-time data on insect numbers and focus on more effective pesticide treatments [57].

5.11.1.4 Sustainable Agriculture Goals

With Semios, farmers can reduce the use of pesticides. AI-driven techniques improve the quality of the crop. The growing customer demand for safer food items can be satisfied with Semios, which simultaneously reduces the overall effect on the environment [58].

5.12 IPM: A HOLISTIC APPROACH TO PEST CONTROL

IPM is a long-term strategy. Here biological, cultural, physical, and chemical approaches are combined. Reduction of economic, health, and environmental hazards makes IPM suitable for eco-friendly homeowners, farms, and eco companies.

5.12.1 BENEFITS OF IPM

IPM reduces pest populations, and also reduces the cases of chemical resistance. Use of non-chemical solutions is the main priority of this method. This method employs chemicals only when there is no other alternative available. As a result, over time it is cost-effective and friendly to the environment.

5.12.2 HOW IPM WORKS?

IPM helps to identify pests. It (1) establishes action thresholds, (2) tracks the activity of pests, and (3) executes management techniques focusing on non-chemical remedies. By evaluating regularly, the plan's effectiveness is assured. Adjustments are applied as necessary [59].

5.12.3 EXAMPLES OF IPM

In IPM tactics, the release of ladybugs controls aphids. Rotational crops reduce the pests to be developed.

5.13 REVOLUTIONIZING PEST CONTROL: WAINS IoT AND AI INNOVATIONS

WAINS is a technology firm that is based in Albstadt, Germany. With a unique combination of IoT technology and picture recognition powered by AI, this technology is redefining the process of pest management. WAINS was founded in late 2019. For the larger population of the world's 8 billion people, guaranteeing a sustainable food supply chain is too challenging. With WAINS, we can address this challenge. This technology manages insects in stored food and enhances the facilities where food is prepared.

5.13.1   THE PROBLEM OF PESTS

Mice and weevils infest the food industry frequently. Benjamin Ruoff is the managing director of WAINS. He has emphasized why it is essential to detect pests in an early stage with the help of digitalization. WAINS targets digitally, creating an early warning system for controlling pests in several industries. These industries also include pharmaceuticals and hotels. They combine Cumulocity IoT with AI-powered picture recognition.

5.13.2   INNOVATIVE SOLUTIONS

WAINS has developed a unique type of pest trap. They include attractant tablets and pheromones that usually entice pests. These traps are equipped with sensors and cameras, which are monitored continuously. They click photographs of the field environment. This data is then transmitted to the cloud. This occurs on a daily basis. Here the Cumulocity IoT first analyzes and then displays it. The analyzed data is displayed in a dashboard that is easy to use. Technically, a comprehensive insect taxonomy database is at the core of this system. Based on the database, pests are identified and classified in an autonomous manner. At the end of the process, the system provides its users with very quick notifications.

5.13.3   EFFICIENCY AND SUSTAINABILITY

Traditional pest management procedures were frequently inspected monthly by hand. This manual inspection was slow and ineffective. Compared to that, WAINS system monitors in real time. They have drastically lowered the time when pests were unnoticed previously. The required chemical interventions have also been lowered. The European Union's Green Deal seeks to halve pesticide use. Their target is 2030. WAINS system is consistent with this deal.

5.13.4   PREDICTIVE PEST CONTROL VISION

WAINS sees a future. Their predictive pest control is enabled by digital monitoring and analysis. This organization documents infestation alongside environmental conditions. This allows them to identify the underlying causes and the possible and efficient target solutions. Their intention is to work with biologists. Those professionals will keep an up-to-date pest database. In that database, rare and invasive species will also be included. In that way, they will improve the capabilities of their system.

5.13.5   DIGITAL TRANSFORMATION IN PEST CONTROL

Traptice is a solution by WAINS. It is a smart pest management gadget. Analog and digital components are combined within this gadget. The system consists of (1) a monitoring box that is stocked with attractants and (2) a digital unit. It collects environmental data and (3) AI-powered analysis that can classify pests that have been trapped. Traptice allows pest management firms to optimize their operations. Those farms can then enable remote monitoring via Wi-Fi or LoRaWAN connections. This helps to cut human expenses.

5.13.6 COST-EFFECTIVE AND FLEXIBLE

WAINS provided Traptices are lease options. They make it available for organizations to test. We can utilize the saved time for improving customer interactions. Preventative measures can be developed against pest infestations. Some clients prefer using biological methods. They target raising beneficial insects that may be conjugated with WAINS technology.

5.13.7 EXPANDING OPPORTUNITIES

WAINS had to overcome obstacles during the COVID-19 pandemic. Despite that, since its establishment, WAINS has doubled in size. Not only are they rapidly expanding, they are also posing their impacts in industrial, pharmaceutical, and hospitality. For controlling pests, WAINS represents a comprehensive data hub. This hub will continually develop as they simultaneously explore niche markets. Such markets include digital bedbug monitoring in hotels [60].

5.14 EMPOWERING COTTON FARMERS WITH CottonAce: AN AI-DRIVEN SOLUTION

Cotton farming supports approximately 100 million farmers all over the world. Of them, 90% are farmers who produce on a small scale. They mainly come from low- and middle-income countries. India is the world's greatest cotton producer. India cultivates 26% of the world's total cotton. The farming spans across 41% of the total area of land. However, these farmers face considerable obstacles in farming. Severe pest infestations are common in cotton cultivation. Among them, the Pink Bollworm (PBW) has become resistant against Bt cotton. This incident of resistance impacts crop losses up to 30% per year. As a result, financial instability becomes a daily companion among farmers.

5.14.1 INTRODUCING CottonAce

CottonAce is an AI-powered system with an early warning feature. While using it, smallholder cotton producers are able to manage insect-associated risks with great efficiency. With an easy-to-use app installed on Android devices, this system recommends farmers apply pesticides from time to time based on their specific conditions.

5.14.2 HOW IT WORKS?

First, pests are caught in pheromone traps. Second, farmers, particularly lead farmers working with welfare programs, upload photographs of those pests to the

CottonAce app. Third, the AI program scrutinizes these photos. They recognize and measure bugs and calculate the infestation levels. However, in a developing country like India, not all farmers may have access to mobile facilities and/or the application. Therefore, lastly, this information is shared with nearby farmers, who cannot access the application.

5.14.3 FEATURES AND IMPACT

Unlike other applications, CottonAce works offline. It supports nine languages. There is a web-based dashboard that can monitor insects in real time. This dashboard eases the process of tracking infestations and helps farmers follow-up. Being AI-focused, this application detects pests with high accuracy. With this accurate detection, farmers only spray pesticides when it is really mandatory, reducing chemical use. An environment-friendly approach becomes the agricultural practice.

CottonAce was first launched in 2021, when the Kharif season was ongoing. And by the end of the next Kharif season of 2023, the company hoped to reach 500,000 farmers. For the early results, we can comment that those farmers who were using this app increased their income. Their quality of yield was remarkably higher than others statistically. This indicates a positive step where pest management in cotton cultivation can be transformed [61].

5.15 NAVIGATING THE CHALLENGES OF AI IN AGRICULTURE

AI revolutionizes agriculture by increasing production, sustainability, and efficiency. However, there are a number of barriers that prevent them from being adopted widely.

(1) *Data accessibility and quality:* The availability and quality of agricultural data is a main obstacle. The data can be diverse, limited, and unreliable. Creation of accurate AI models needs data on weather, soil conditions, crop health, and historical yields on which we can rely. Unfortunately, there are many rural communities that don't have the infrastructure that is required for time-to-time data collection.

(2) *Complexity of agricultural systems:* The agricultural systems are intrinsically complex. This makes AI integration difficult. Weather, qualities of soil, crop genetics, and pest dynamics are interconnected factors that we need to consider. These factors necessitate specific solutions based on the diverse geographies and their particular cropping systems.

(3) *Transparency and interpretability:* Farmers demand comprehensive explanations when relying on AI-given advice. The black-box nature of the several AI algorithms is not explainable. Here consumers withdraw their trust. To receive approval from users, the interpretability of those models needs to ensured.

(4) *Affordability and accessibility:* When there is a case of resource-constrained contexts, especially among smallholder farmers accessing AI tools can be limited in rural areas. Data privacy and security are other problems including demand for protection against illegal access.

(5) *Robustness and scalability:* To work effectively, AI models have to be adaptable according to changes in climate. The scalability should be acquired across different farm sizes and types of crops. Good communication, training, and extension services are needed to overcome skepticism among farmers and resistance among institutions [62].

5.15.1 Overcoming Adoption Bottlenecks in AI Agriculture

In agriculture, the effectiveness of smart systems can be determined as per their ability to complete a task swiftly and with higher precision. Unfortunately, there are some systems that sluggish reaction times and errors, disrupt user processes. This occurs as a result of (1) accessibility to balancing and (2) accuracy of the output.

(1) Automation, (2) speed, and (3) capability to monitor – these are the three essential tactics that improve the performance of AI. To create an agricultural intelligence system and ensure its success, collaboration is needed not only among experts from various fields of agriculture but also among the farmers who are actually going to utilize it.

Most AI systems are web-based. Thereafter, farmers from isolated and rural areas can suffer from accessibility issues. To help those farmers, governments can develop affordable web services. These systems will engage those farmers directly with AI systems. Training programs can also assist farmers to become familiar with these technologies.

To be successful, AI systems must be adaptive. They should have the capability to be integrated into a larger ecosystem. This will allow them to absorb a variety of inputs from its users, which can be used to improve performance [63].

5.15.2 The Promising Horizon of AI in Agriculture

Using AI in agriculture has already delivered considerable amounts of benefits. It is expected that future breakthroughs will expand its applicability. BI Intelligence Research claims and predicts that the market for agricultural technology, like AI and ML, will be treble by 2025. It will reach $15.3 billion [64]. AI and ML are sophisticated analytical technologies. They use historical data. Then they foresee and reduce damage caused by insects. It increases the output. Future advancements will enhance the use of autonomous machines and drones. They will be employed for planting, fertilizing, weeding, and harvesting. Human workload will be decreased. Efficiency will be increased. AI will also offer predictive data on pest and climate risks to farmers. These will help them to practice climate-smart farming. They will be tailored to manage their water, select their crops, and control the management of pests.

We have already seen some innovative partnerships, like those between Microsoft and the International Crop Research Institute for Semi-Arid Tropics (ICRISAT). These partnerships will transform techniques of planting. Recommendations based on AI will boost output by 30%. The aim of the collaborative initiatives with United Phosphorus Limited (UPL) is to provide an API that will forecast pest-associated risks using AI for pest invasion prediction [65].

The life cycles of some pests are complex. Some pests live underground. Research on pest detection is still restricted. For pest identification, most of the research has only focused on CNNs, the traditional ones. This traditional technique needs frequent substantial training data and demands processing resources. Future studies should focus on CNN designs that are more compact. MobileNet and SqueezeNet are types of compact designs which are increasing efficiency and lowering costs [66, 67]. Furthermore, as in many agricultural contexts, previous studies have not delivered a clear idea about the dynamics of insect populations. How abiotic factors impact pest behavior and the performance of AI have to be explored further.

5.16 CONCLUSIVE REMARKS

In the context of the agricultural industry, integrating AI in smart pest management is a game-changing possibility. Particularly, this is important for smallholder producers who deal with major issues of pest control and crop productivity. In this chapter, it has been shown that AI technologies, including image recognition, ML, and predictive analytics, have possibilities to address the pest dynamics complexities successfully. The processes of decision-making processes are improving. Sustainability of overall farming will increase. AI-driven solutions like CottonAce delivered positive outcomes. It improved crop yields, reduced the use of harmful pesticides, and ensured that farmers could make more profits. With these innovations, not only economic resilience was improved, but also environment-friendly methods were promoted. To ensure global food security is the priority of these methods. Accessibility of data, model interpretability, and the need for personalized solutions accounting for varying types and conditions of farming systems are the obstacles that are needed to be overcome to realize the full potential of AI in agricultural fields. Future research should improve the adaptability of AI models. We have to expand the outreach to resource-constrained farmers. The advantages of AI have to be delivered equally across varied agricultural settings. Collaboration among developers, stakeholders in agriculture, and farmers should explore AI's potential that drives innovation and supports sustainable practices in farming. By continuous investment in research and education, we can pave a road that is more efficient and resilient in the agricultural sector and have the potential to thrive in the face of challenges that change rapidly.

AUTHOR CONTRIBUTIONS

Both the authors contributed equally to prepare the chapter and approved the submitted version.

ACKNOWLEDGMENTS

The author (S. Roy) would like to acknowledge "Scheme for Transformational and Advanced Research in Sciences" (STARS) (MoE-STARS/STARS-2/2023–0175) by the Ministry of Education for promoting translational India-centric research in sciences implemented and managed by Indian Institute of Science (IISc), Bangalore, for their support.

REFERENCES

[1] N. Ahmed, D. De, and I. Hussain, "Internet of Things (IoT) for Smart Precision Agriculture and Farming in Rural Areas," *IEEE Internet of Things Journal*, vol. 5, no. 6, pp. 4890–4899, 2018, https://doi.org/10.1109/JIOT.2018.2879579.

[2] W. Li, D. Wang, M. Li, Y. Gao, J. Wu, and X. Yang, "Field Detection of Tiny Pests from Sticky Trap Images Using Deep Learning in Agricultural Greenhouse," *Computers and Electronics in Agriculture*, vol. 183, p. 106048, 2021, https://doi.org/10.1016/j.compag.2021.106048.

[3] M. K. Boutahir, A. O. Alaoui, Y. Farhaoui, M. Azrour, and A. E. Allaoui, "Efficient Solar Radiation Prediction through Adaptive Neighborhood Rough Set-Based Feature Selection in Meteorological Streaming Data in Errachidia, Morocco," in *Internet of Things and Big Data Analytics for a Green Environment*, Chapman & Hall/CRC, 2024.

[4] N. Ben-Lhachemi, M. Benchrifa, S. Nasrdine, J. Mabrouki, M. Slaoui, and M. Ade Azrour, "Effect of IoT Integration in Agricultural Greenhouses," in *Technical and Technological Solutions Towards a Sustainable Society and Circular Economy*, J. Mabrouki and A. Mourade, Eds., Cham: Springer Nature Switzerland, 2024, pp. 435–445, https://doi.org/10.1007/978-3-031-56292-1_35.

[5] W. Shafik and M. Azrour, "Building a Greener World: Harnessing the Power of IoT and Smart Devices for Sustainable Environment," in *Technical and Technological Solutions Towards a Sustainable Society and Circular Economy*, J. Mabrouki and A. Mourade, Eds., Cham: Springer Nature Switzerland, 2024, pp. 35–58, https://doi.org/10.1007/978-3-031-56292-1_3.

[6] S. Dargaoui, et al., "IoT-Driven Smart Agriculture: Security Issues and Authentication Schemes Classification," in *Proceeding of the International Conference on Connected Objects and Artificial Intelligence (COCIA2024)*, Y. Mejdoub and A. Elamri, Eds., Cham: Springer Nature Switzerland, 2024, pp. 61–66, https://doi.org/10.1007/978-3-031-70411-6_10.

[7] K. Bella, et al., "An Efficient Intrusion Detection System for IoT Security Using CNN Decision Forest," *PeerJ Computer Science*, vol. 10, p. e2290, 2024, https://doi.org/10.7717/peerj-cs.2290.

[8] V. Partel, S. Charan Kakarla, and Y. Ampatzidis, "Development and Evaluation of a Low-Cost and Smart Technology for Precision Weed Management Utilizing Artificial Intelligence," *Computers and Electronics in Agriculture*, vol. 157, pp. 339–350, 2019, https://doi.org/10.1016/j.compag.2018.12.048.

[9] M. Azrour, et al., "A Survey of Machine and Deep Learning Applications in the Assessment of Water Quality," in *World Sustainability Series*, vol. Part F2854, 2024, pp. 471–483, https://doi.org/10.1007/978-3-031-56292-1_38.

[10] H. Hissou, S. Benkirane, A. Guezzaz, A. Beni-Hssane, and M. Azrour, "Advanced Prediction of Solar Radiation Using Machine Learning and Principal Component Analysis," in *Artificial Intelligence, Data Science and Applications*, Y. Farhaoui, A. Hussain, T. Saba, H. Taherdoost, and A. Verma, Eds., Cham: Springer Nature Switzerland, 2024, pp. 201–207, https://doi.org/10.1007/978-3-031-48573-2_29.

[11] V. Dhananjayan and B. Ravichandran, "Occupational Health Risk of Farmers Exposed to Pesticides in Agricultural Activities," *Current Opinion in Environmental Science & Health*, vol. 4, pp. 31–37, 2018, https://doi.org/10.1016/j.coesh.2018.07.005.

[12] A. V. A. Machado, D. M. Potin, J. B. Torres, and C. S. A. Silva Torres, "Selective Insecticides Secure Natural Enemies Action in Cotton Pest Management," *Ecotoxicology and Environmental Safety*, vol. 184, p. 109669, 2019, https://doi.org/10.1016/j.ecoenv.2019.109669.

[13] N. Shah and S. Jain, "Detection of Disease in Cotton Leaf Using Artificial Neural Network," in *2019 Amity International Conference on Artificial Intelligence (AICAI)*, 2019, pp. 473–476, https://doi.org/10.1109/AICAI.2019.8701311.

[14] M. Calvo-Agudo, J. F. Tooker, M. Dicke, and A. Tena, "Insecticide-Contaminated Honeydew: Risks for Beneficial Insects," *Biological Reviews*, vol. 97, no. 2, pp. 664–678, 2022, https://doi.org/10.1111/brv.12817.

[15] S. Van Goethem, S. Verwulgen, F. Goethijn, and J. Steckel, "An IoT Solution for Measuring Bee Pollination Efficacy," in *2019 IEEE 5th World Forum on Internet of Things (WF-IoT)*, 2019, pp. 837–841, https://doi.org/10.1109/WF-IoT.2019.8767298.

[16] P. Wang, Z. Zhu, Q. Chen, and W. Dai, "Text Reasoning Chain Extraction for Multi-Hop Question Answering," *Tsinghua Science and Technology*, vol. 29, no. 4, pp. 959–970, 2024, https://doi.org/10.26599/TST.2023.9010060.

[17] G. G. and A. P. J., "Identification of Plant Leaf Diseases Using a Nine-Layer Deep Convolutional Neural Network," *Computers & Electrical Engineering*, vol. 76, pp. 323–338, 2019, https://doi.org/10.1016/j.compeleceng.2019.04.011.

[18] P. Boissard, V. Martin, and S. Moisan, "A Cognitive Vision Approach to Early Pest Detection in Greenhouse Crops," *Computers and Electronics in Agriculture*, vol. 62, no. 2, pp. 81–93, 2008, https://doi.org/10.1016/j.compag.2007.11.009.

[19] G. L. Tenório, et al., "Comparative Study of Computer Vision Models for Insect Pest Identification in Complex Backgrounds," in *2019 12th International Conference on Developments in eSystems Engineering (DeSE)*, 2019, pp. 551–556, https://doi.org/10.1109/DeSE.2019.00106.

[20] R. Lahlali, et al., "Biological Control of Plant Pathogens: A Global Perspective," *Microorganisms*, vol. 10, no. 3, Art. no. 3, 2022, https://doi.org/10.3390/microorganisms10030596.

[21] A. X. Silva, G. Jander, H. Samaniego, J. S. Ramsey, and C. C. Figueroa, "Insecticide Resistance Mechanisms in the Green Peach Aphid Myzus Persicae (Hemiptera: Aphididae) I: A Transcriptomic Survey," *PLoS ONE*, vol. 7, no. 6, p. e36366, 2012, https://doi.org/10.1371/journal.pone.0036366.

[22] G. V. P. Reddy and A. Guerrero, "Chapter Twenty – New Pheromones and Insect Control Strategies," in *Vitamins & Hormones*, G. Litwack, Ed., in Pheromones, vol. 83, Academic Press, 2010, pp. 493–519, https://doi.org/10.1016/S0083-6729(10)83020-1.

[23] S. Chinta, R. Vander Meer, E. O'Reilly, and M.-Y. Choi, "Insecticidal Effects of Receptor-Interference Isolated Bioactive Peptides on Fire Ant Colonies," *International Journal of Molecular Sciences*, vol. 24, no. 18, Art. no. 18, 2023, https://doi.org/10.3390/ijms241813978.

[24] B. Manachini, P. Casati, L. Cinanni, and P. Bianco, "Role of Myzus Persicae (Hemiptera: Aphididae) and Its Secondary Hosts in Plum Pox Virus Propagation," *Journal of Economic Entomology*, vol. 100, no. 4, pp. 1047–1052, 2007, https://doi.org/10.1093/jee/100.4.1047.

[25] C. Concha, et al., "A Transgenic Male-Only Strain of the New World Screwworm for an Improved Control Program Using the Sterile Insect Technique," *BMC Biology*, vol. 14, no. 1, p. 72, 2016, https://doi.org/10.1186/s12915-016-0296-8.

[26] J. Zhang, F. Kong, J. Wu, S. Han, and Z. Zhai, "Automatic Image Segmentation Method for Cotton Leaves with Disease Under Natural Environment," *Journal of Integrative Agriculture*, vol. 17, no. 8, pp. 1800–1814, 2018, https://doi.org/10.1016/S2095-3119(18)61915-X.

[27] R. Fjelland, "Why General Artificial Intelligence Will Not Be Realized," *Humanities and Social Sciences Communications*, vol. 7, no. 1, pp. 1–9, 2020, https://doi.org/10.1057/s41599-020-0494-4.

[28] M. Vitek and P. Peer, "Chapter Five – Intelligent Agents in Games: Review with an Open-Source Tool," in *Advances in Computers*, vol. 116, 1 vols., A. R. Hurson and V. Milutinović, Eds., Elsevier, 2020, pp. 251–303, https://doi.org/10.1016/bs.adcom.2019.07.005.

[29] S. Kar, et al., "Self-Supervised Learning Improves Classification of Agriculturally Important Insect Pests in Plants," *The Plant Phenome Journal*, vol. 6, no. 1, p. e20079, 2023, https://doi.org/10.1002/ppj2.20079.

[30] M. Wakchaure, B. K. Patle, and A. K. Mahindrakar, "Application of AI Techniques and Robotics in Agriculture: A Review," *Artificial Intelligence in the Life Sciences*, vol. 3, p. 100057, 2023, https://doi.org/10.1016/j.ailsci.2023.100057.

[31] J. G. A. Barbedo, "Using Digital Image Processing for Counting Whiteflies on Soybean Leaves," *Journal of Asia-Pacific Entomology*, vol. 17, no. 4, pp. 685–694, 2014, https://doi.org/10.1016/j.aspen.2014.06.014.

[32] I. H. Sarker, "Deep Learning: A Comprehensive Overview on Techniques, Taxonomy, Applications and Research Directions," *SN Computer Science*, vol. 2, no. 6, p. 420, 2021, https://doi.org/10.1007/s42979-021-00815-1.

[33] H. Hissou, S. Benkirane, A. Guezzaz, M. Azrour, and A. Beni-Hssane, "A Novel Machine Learning Approach for Solar Radiation Estimation," *Sustainability*, vol. 15, no. 13, Art. no. 13, 2023, https://doi.org/10.3390/su151310609.

[34] H. Attou, A. Guezzaz, S. Benkirane, M. Azrour, and Y. Farhaoui, "Cloud-Based Intrusion Detection Approach Using Machine Learning Techniques," *Big Data Mining and Analytics*, vol. 6, no. 3, pp. 311–320, 2023.

[35] L. Deng, Y. Wang, Z. Han, and R. Yu, "Research on Insect Pest Image Detection and Recognition Based on Bio-Inspired Methods," *Biosystems Engineering*, vol. 169, pp. 139–148, 2018, https://doi.org/10.1016/j.biosystemseng.2018.02.008.

[36] S. A. Shah, et al., "Application of Drone Surveillance for Advance Agriculture Monitoring by Android Application Using Convolution Neural Network," *Agronomy*, vol. 13, no. 7, Art. no. 7, 2023, https://doi.org/10.3390/agronomy13071764.

[37] J. L. Blackall, J. Wang, M. R. A. Nabawy, M. K. Quinn, and B. D. Grieve, "Development of a Passive Spore Sampler for Capture Enhancement of Airborne Crop Pathogens," *Fluids*, vol. 5, no. 2, Art. no. 2, 2020, https://doi.org/10.3390/fluids5020097.

[38] B. Kariyanna and M. Sowjanya, "Unravelling the Use of Artificial Intelligence in Management of Insect Pests," *Smart Agricultural Technology*, vol. 8, p. 100517, 2024, https://doi.org/10.1016/j.atech.2024.100517.

[39] B. J. Spiesman, et al., "Assessing the Potential for Deep Learning and Computer Vision to Identify Bumble Bee Species from Images," *Scientific Reports*, vol. 11, no. 1, p. 7580, 2021, https://doi.org/10.1038/s41598-021-87210-1.

[40] A. Rajab, et al., "Flood Forecasting by Using Machine Learning: A Study Leveraging Historic Climatic Records of Bangladesh," *Water*, vol. 15, no. 22, Art. no. 22, 2023, https://doi.org/10.3390/w15223970.

[41] X. Zhang, J. Bu, X. Zhou, and X. Wang, "Automatic Pest Identification System in the Greenhouse Based on Deep Learning and Machine Vision," *Frontiers in Plant Science*, vol. 14, 2023, https://doi.org/10.3389/fpls.2023.1255719.

[42] J. Rane, Ö. Kaya, S. K. Mallick, and N. L. Rane, *Generative Artificial Intelligence in Agriculture, Education, and Business*, Deep Science Publishing, 2024.

[43] M. Tannous, C. Stefanini, and D. Romano, "A Deep-Learning-Based Detection Approach for the Identification of Insect Species of Economic Importance," *Insects*, vol. 14, no. 2, Art. no. 2, 2023, https://doi.org/10.3390/insects14020148.

[44] R. K. Dhanaraj and M. A. Ali, "Deep Learning-Enabled Pest Detection System Using Sound Analytics in the Internet of Agricultural Things," *Engineering Proceedings*, vol. 58, no. 1, Art. no. 1, 2023, https://doi.org/10.3390/ecsa-10-16205.

[45] R. Prabha, J. S. Kennedy, G. Vanitha, N. Sathiah, and M. B. Priya, "Android Application Development for Identifying Maize Infested with Fall Armyworms with Tamil Nadu Agricultural University Integrated Proposed Pest Management (TNAU IPM) Capsules," *Journal of Applied and Natural Science*, vol. 14, no. SI, Art. no. SI, 2022, https://doi.org/10.31018/jans.v14iSI.3599.

[46] R. Sharma, "Artificial Intelligence in Agriculture: A Review," in *2021 5th International Conference on Intelligent Computing and Control Systems (ICICCS)*, IEEE, 2021, pp. 937–942.

[47] R. Toscano-Miranda, M. Toro, J. Aguilar, M. Caro, A. Marulanda, and A. Trebilcok, "Artificial-Intelligence and Sensing Techniques for the Management of Insect Pests and Diseases in Cotton: A Systematic Literature Review," *Journal of Agricultural Science*, vol. 160, no. 1–2, pp. 16–31, 2022, https://doi.org/10.1017/S002185962200017X.

[48] A. Khattab, S. E. D. Habib, H. Ismail, S. Zayan, Y. Fahmy, and M. M. Khairy, "An IoT-Based Cognitive Monitoring System for Early Plant Disease Forecast," *Computers and Electronics in Agriculture*, vol. 166, p. 105028, 2019, https://doi.org/10.1016/j.compag.2019.105028.

[49] A. N. Brezolin, et al., "Tools for Detecting Insect Semiochemicals: A Review," *Analytical and Bioanalytical Chemistry*, vol. 410, no. 17, pp. 4091–4108, 2018, https://doi.org/10.1007/s00216-018-1118-3.

[50] E. Biondi, et al., "Detection of Potato Brown Rot and Ring Rot by Electronic Nose: From Laboratory to Real Scale," *Talanta*, vol. 129, pp. 422–430, 2014, https://doi.org/10.1016/j.talanta.2014.04.057.

[51] S. Xu, et al., "Estimation of the Age and Amount of Brown Rice Plant Hoppers Based on Bionic Electronic Nose Use," *Sensors*, vol. 14, no. 10, Art. no. 10, 2014, https://doi.org/10.3390/s141018114.

[52] T. Wen, et al., "Rapid Detection and Classification of Citrus Fruits Infestation by *Bactrocera dorsalis* (Hendel) Based on Electronic Nose," *Postharvest Biology and Technology*, vol. 147, pp. 156–165, 2019, https://doi.org/10.1016/j.postharvbio.2018.09.017.

[53] A. Hafeez, et al., "Implementation of Drone Technology for Farm Monitoring & Pesticide Spraying: A Review," *Information Processing in Agriculture*, vol. 10, no. 2, pp. 192–203, 2023, https://doi.org/10.1016/j.inpa.2022.02.002.

[54] M. A. Alahe, et al., "Cyber Security in Smart Agriculture: Threat Types, Current Status, and Future Trends," *Computers and Electronics in Agriculture*, vol. 226, p. 109401, 2024, https://doi.org/10.1016/j.compag.2024.109401.

[55] D. Mhlanga, "Digital Transformation of the Agricultural Industry in Africa," in *Fostering Long-Term Sustainable Development in Africa: Overcoming Poverty, Inequality, and Unemployment*, D. Mhlanga and M. Dzingirai, Eds., Cham: Springer Nature Switzerland, 2024, pp. 441–464, https://doi.org/10.1007/978-3-031-61321-0_19.

[56] D. O. Kiobia, C. J. Mwitta, K. G. Fue, J. M. Schmidt, D. G. Riley, and G. C. Rains, "A Review of Successes and Impeding Challenges of IoT-Based Insect Pest Detection Systems for Estimating Agroecosystem Health and Productivity of Cotton," *Sensors*, vol. 23, no. 8, Art. no. 8, 2023, https://doi.org/10.3390/s23084127.

[57] A. K. Singh, S. Kumar, and B. Jyoti, "Influence of Climate Change on Agricultural Sustainability in India: A State-Wise Panel Data Analysis," *Asian Journal of Agriculture*, vol. 6, no. 1, Art. no. 1, 2022, https://doi.org/10.13057/asianjagric/g060103.

[58] M. Lenox and R. Duff, *The Decarbonization Imperative: Transforming the Global Economy by 2050*, Stanford University Press, 2021.

[59] K. Karlsson Green, J. A. Stenberg, and Å. Lankinen, "Making Sense of Integrated Pest Management (IPM) in the Light of Evolution," *Evolutionary Applications*, vol. 13, no. 8, pp. 1791–1805, 2020, https://doi.org/10.1111/eva.13067.

[60] B. Paul, M. Paul, and A. Rub, "Advancements in AI-Based Pest and Disease Detection in Agriculture: A Comprehensive Review of Image Recognition and Disease Modelling," *Social Science Research Network, Rochester, NY,* p. 4843850, 2024, https://doi.org/10.2139/ssrn.4843850.

[61] D. Zhang, Y. Zhang, L. Sun, J. Dai, and H. Dong, "Mitigating Salinity Stress and Improving Cotton Productivity with Agronomic Practices," *Agronomy,* vol. 13, no. 10, Art. no. 10, 2023, https://doi.org/10.3390/agronomy13102486.

[62] T. Talaviya, D. Shah, N. Patel, H. Yagnik, and M. Shah, "Implementation of Artificial Intelligence in Agriculture for Optimisation of Irrigation and Application of Pesticides and Herbicides," *Artificial Intelligence in Agriculture,* vol. 4, pp. 58–73, 2020, https://doi.org/10.1016/j.aiia.2020.04.002.

[63] I. Gryshova, et al., "Artificial Intelligence in Climate Smart in Agricultural: Toward a Sustainable Farming Future," *ACCESS,* vol. 5, no. 1, pp. 125–140, 2024, https://doi.org/10.46656/access.2024.5.1(8).

[64] Rimpika, et al., "An Overview of Precision Farming," *International Journal of Environment and Climate Change,* vol. 13, no. 12, Art. no. 12, 2023, https://doi.org/10.9734/ijecc/2023/v13i123701.

[65] M. Javaid, A. Haleem, I. H. Khan, and R. Suman, "Understanding the Potential Applications of Artificial Intelligence in Agriculture Sector," *Advanced Agrochem,* vol. 2, no. 1, pp. 15–30, 2023, https://doi.org/10.1016/j.aac.2022.10.001.

[66] P. V. Dantas, W. Sabino da Silva, L. C. Cordeiro, and C. B. Carvalho, "A Comprehensive Review of Model Compression Techniques in Machine Learning," *Applied Intelligence,* vol. 54, no. 22, pp. 11804–11844, 2024, https://doi.org/10.1007/s10489-024-05747-w.

[67] A. Dembele, R. W. Mwangi, and A. O. Kube, "A Lightweight Convolutional Neural Network with Hierarchical Multi-Scale Feature Fusion for Image Classification," *Journal of Computer and Communications,* vol. 12, no. 2, Art. no. 2, 2024, https://doi.org/10.4236/jcc.2024.122011.

6 Artificial Intelligence
Technological Revolution for Moroccan Agriculture

Hajar Khabouche, Miloudia Slaoui, and Jamal Mabrouki

6.1 INTRODUCTION

The economic, social, and geographical challenges facing the agricultural sector have led Morocco to prioritize its development choices since independence. Over the years, the sector's impact on the national economy has become increasingly visible, both through its own performance and through its interaction with other sectors. This is evidenced, for example, by its contribution to GDP (around 12%) and its role in providing employment for a significant proportion of the population (around 38% of the workforce), not to mention the critical role the sector plays in several regions of the Kingdom [1, 2].

Due to frequent droughts and erratic rainfall, Morocco is vulnerable to climate change, which has a negative impact on rural yields and water accessibility. Agricultural productivity is also limited by soil degradation and ineffective water resource management. The inefficient supply chain and low adoption of current technologies exacerbate postharvest losses and affect profitability [3]. Finally, limited access to markets and finance prevents farmers from investing in more productive and sustainable practices [4].

The use of artificial intelligence (AI) to create intelligent processes – including techniques and precision agriculture [5–9] – is also beginning to take shape. In the future, agriculture will not be limited to conventional techniques such as irrigation, fertilizers, and seedlings. Digital platforms and the use of algorithms will also have an impact. The world will be affected by this digital revolution, which is gradually spreading across the African continent.

This chapter highlights the importance of creating a coordinated approach that balances natural conservation, social well-being, and financial progress by identifying the challenges and opportunities that lie ahead. This integrative strategy is essential as it recognizes the interconnectedness of ecological health, community prosperity, and economic sustainability.

As the world faces unprecedented environmental changes, it becomes increasingly clear that conservation efforts cannot succeed in isolation. Thus, we must actively engage various stakeholders, including local communities, businesses, and

DOI: 10.1201/9781003527664-6

policymakers, to develop solutions that are not only environmentally sound but also socially equitable and economically viable. This coordinated approach encourages collaborative efforts that leverage diverse perspectives and resources, fostering innovation and resilience in the face of impending challenges.

Furthermore, this chapter delves into specific challenges, such as habitat loss, climate change, and socioeconomic disparities, while also exploring the opportunities presented by emerging technologies, community engagement, and policy reforms. By addressing these critical issues holistically, we can create frameworks that promote sustainable development, protect natural ecosystems, and enhance the quality of life for all individuals.

6.2   OVERVIEW

Artificial intelligence (AI) is an exciting field of computer science that enables the creation of systems capable of performing tasks that typically require human intelligence. These systems use algorithms and mathematical models to learn from data, recognize patterns, and make decisions on their own [10, 11]. The term has undergone significant evolution over time, currently encompassing all ideas aimed at enabling a machine to emulate and surpass human cognitive abilities. The term "artificial intelligence" emerged in 1956, following a surge of studies after World War II, and represents one of the youngest research fields in the natural sciences and engineering. This followed numerous inventions made by machines over the centuries. The hypothesis arose that it is possible to compute and that machines can think and act independently [6, 12]. John McCarthy is credited with coining the term "artificial intelligence," which is frequently shortened to "AI." "The science and engineering of creating intelligent machines, especially intelligent computer programs," is how McCarthy describes artificial intelligence. Although it is comparable to using computers to study human intellect, artificial intelligence shouldn't be restricted to techniques that can be observed in biology. Creating realistic visuals is frequently the goal of these. Generative AI has a wide range of uses, from producing text and images to writing computer programs and improving data [12–14].

Artificial intelligence (AI) and the Internet of Things (IoT) have been introduced by agricultural farms that raise crops, fish, and cattle in an effort to increase output and quality [15, 16]. Robotics and artificial intelligence technology can be widely applied throughout the agricultural value chain, from the farm to the consumer, to successfully handle the difficulties facing the industry [7]. The future of agriculture and the farming industry depends on the development of innovative concepts and technological solutions that can enhance productivity and improve the utilization of resources, with the aid of unconventional computing tools. The progressive utilization of crop models and decision-making tools in the agricultural field has the potential to enhance production and resource utilization efficiency. There is a significant opportunity for artificial intelligence to revolutionized agriculture by integrating advanced technologies to forecast agricultural productivity [6, 8, 17]. The application of unmanned aerial vehicle (UAV) and satellite technology to the field of remote sensing. Artificial intelligence (AI) and the Internet of Things (IoT) are being adopted by farms that raise animals, grow crops, and raise aquaculture

[18]. The application of artificial intelligence in real-time data monitoring has demonstrated considerable value in the management of crops, pests, weeds, and yields. Machines communicate with one another in order to ascertain the optimal crops for marketing and harvesting. This provides farmers with effective field management strategies and healthier crops. Artificial intelligence facilitates the delivery of timely information through the appropriate channels, thereby assisting users in developing resilience [19].

6.3 MATERIALS AND METHODS

In Morocco, a variety of artificial intelligence (AI) technologies and applications are employed to modernize and enhance agricultural productivity. These include several types of sensors that monitor significant environmental factors, which are integrated into Internet of Things (IoT) sensor systems utilized in the agricultural sector. The humidity sensor in the soil, for instance, measures the soil's moisture content and provides valuable information for determining the plants' irrigation needs. Temperature sensors are capable of identifying fluctuations in ambient temperature and establishing optimal conditions for plant growth [20, 21] By measuring atmospheric gas concentrations, such as carbon dioxide and other oxides, air quality sensors facilitate a more comprehensive understanding of environmental factors influencing plant growth. To guarantee precise and uninterrupted data collection, the sensors are secured on structures in the fields or directly embedded in the soil. The data is transmitted across wireless networks, enabling it to be sent to centralized management platforms. The aforementioned platforms analyze the data collected to provide recommendations regarding irrigation and fertilization schedules. The real-time analysis enables the modification of agricultural practices, either manually or automatically, in order to meet the specific requirements of a diverse range of cultural contexts. Figure 6.1 provides an overview of the functioning of IoT sensor systems in the agricultural sector, with a particular focus on the monitoring and management of critical environmental factors.

The majority of agricultural drones are multicolor models equipped with high-definition cameras and multispectral sensors. These tools facilitate accurate monitoring of crops and other agricultural products. An integrated GPS system enables precise terrain mapping and navigation, thus facilitating detailed operational management (Figure 6.2). At regular intervals, drones fly over the fields to collect data and take photographs. These images are used to identify and address any anomalies, such as diseases, ravager infestations, and growth problems. The drone also produces a sanitary plant card, which is a valuable tool for effective and focused land management [22, 23].

The disease and predator detection system: Image analysis (IA)-based applications are presently being developed with the aim of identifying plant disease and pest symptoms from plant photographs. These systems make it easier to identify new problems early and take swift action [24]

Online agricultural management platforms: These platforms integrate a multitude of IA technologies, thereby offering comprehensive agricultural management

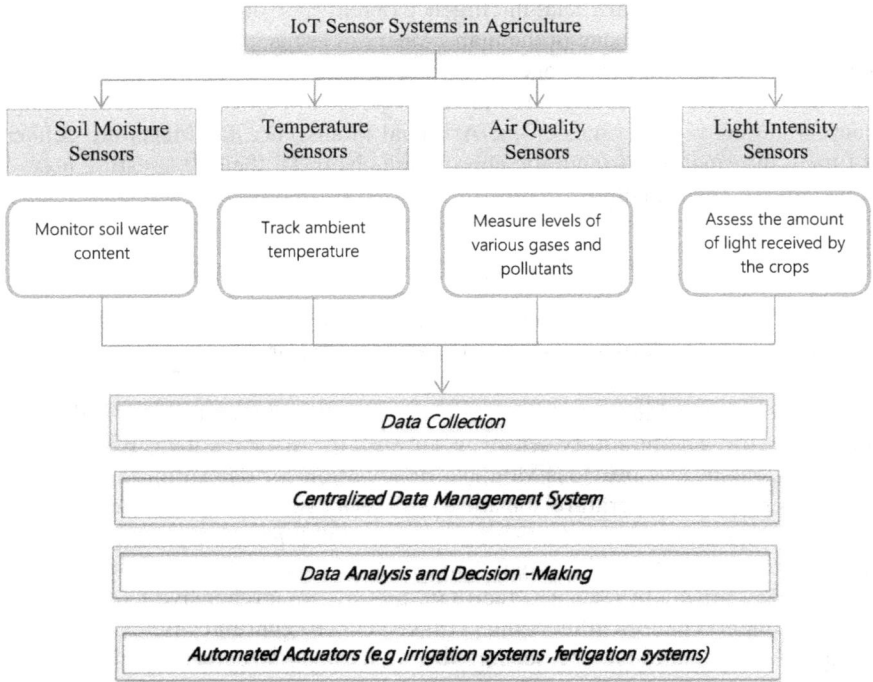

FIGURE 6.1   IoT sensor systems for optimizing agriculture.

FIGURE 6.2   Drone presentation image.

solutions. This enables farmers to effectively plan their operations, manage their resources, and monitor the success of their crops [25].

The deployment of these technologies is expected to enhance the sustainability of agricultural practices, reduce operational costs, and facilitate more precise and effective management of Morocco's agricultural resources.

Table 6.1 outlines the benefits and drawbacks of the aforementioned technologies (drones, online farm management platforms, disease and pest detection systems, and IoT sensor systems) in the agricultural industry.

TABLE 6.1
Advantages and Limitations of Advanced Technologies in Agriculture

Technology	Advantages	Disadvantages
IoT sensor systems	Continuous and precise monitoring of environmental factors (such as air quality, temperature, and soil humidity) Data is transferred in real time to the management platform Adjusting agricultural practices manually or automatically based on data analysis	Initial cost for setting up and maintaining the sensors The reliance on unreliable networks for data transfer The sensors' sensitivity to extreme environmental conditions
Drones	Precise cultural surveillance made possible by high-resolution cameras and multispectral sensors. Accurate mapping and easier navigation, thanks to the integrated GPS Early detection of diseases, pests, and growth issues The creation of plant health cards for targeted management	High initial cost for purchasing drones and related equipment Technical skills are required to operate and maintain drones Restrictions pertaining to weather conditions and battery autonomy
Disease and pest detection systems	Using image analysis to identify pests and disease symptoms early on Quick action to stop the spread of illnesses Decreased application of pesticides via focused identification	– Reliance on the caliber and clarity of the photos that are taken Extensive databases are required in order to accurately identify pests and illnesses The price of creating and putting these procedures into place
Online agricultural management platforms	Combining several AI technologies to provide all encompassing farm management solutions Planning operations and managing resources effectively Tracking crop performance and making wise decisions	Dependence on a reliable Internet connection Farmers must be trained to use these platforms Risks to data security

6.4 RESULTS AND DISCUSSION

To illustrate, the Moroccan enterprise AgriEdge employs these sensors to enhance the efficiency of irrigation in tomato crops by modulating water applications in accordance with the specific soil requirements identified by the sensors [26].

Drones are used by the Moroccan business DR Stone to improve operational effectiveness, reduce harvest losses, and provide vital data for data-driven decision-making (Figure 6.3). The agricultural sector is undergoing a radical change as

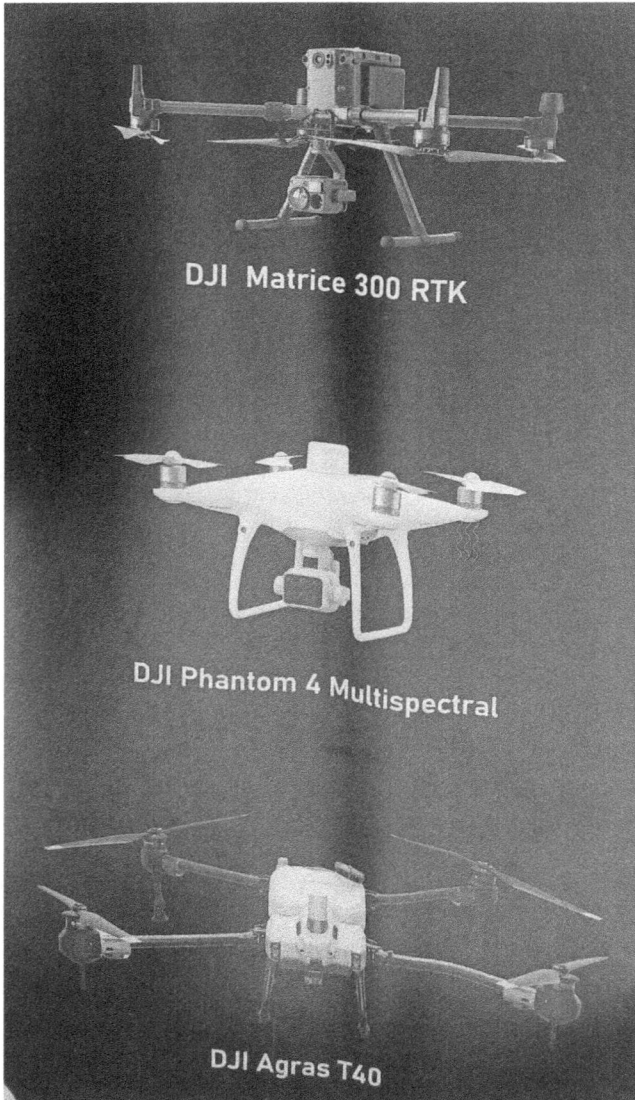

FIGURE 6.3   Types of drones used by the DR Stone organization.

a result of drone integration. An example of this is the use of high-resolution aerial photos to assess crop health. The DJI T50 and T40 drones, data analysis programs like Pix4D Field and Airinov, and pulverization drones like the DJI Agras MG-1S all are part of the company's whole system. In addition to their speed and accuracy, drones are inexpensive and simple to maintain, which lowers operating risks and expenses while enabling operators to access hard-to-reach places. This state-of-the-art technology ensures effective and superior management of agricultural projects by optimizing irrigation, conversion, culture testing, and environmental monitoring [27, 28].

Founded in 2022, DeepLeaf has been at the forefront of AI-based agricultural solutions, transforming farming practices in Morocco and globally. With its headquarters in Technopark Sousse Massa, DeepLeaf cultivates a spirit of innovation and passion that drives the development of creative, tech-enabled solutions for agriculture. The DeepLeaf experimental farm in Sidi Benour, near El Jadida, serves as a "living lab" where the start-up tests its advanced technologies in real-world agricultural conditions. One of DeepLeaf's standout innovations is *Morshida*, an intelligent agricultural assistant accessible via WhatsApp and online platforms. Powered by advanced technologies like deep learning and Natural Language Processing (NLP), Morshida provides farmers with easy access to sophisticated tools, eliminating the need for technical expertise. A key feature of Morshida is its precise identification of various plant diseases through image analysis, enabling farmers to implement preventive measures to protect their crops and enhance productivity. Morshida also utilizes the "DeepLeafLang" model, delivering personalized responses to farmers' questions in Darija, English, and French. Additionally, it integrates real-time weather data to provide location-specific guidance. Another flagship product, *Agrigo*, is a 100% Moroccan-built robot equipped with multispectral cameras to detect plant diseases, further enhancing DeepLeaf's toolkit for modern agriculture.

This innovative robot is capable of accurately and promptly analyzing cultures, thereby enabling the effective prevention of disease. Powered by artificial intelligence and automatic learning, Agrigo has an autonomous pilot system that manages its on-the-ground navigation and allows it to make autonomous decisions about the targeted application of pesticides. This approach has the potential to reduce the excessive use of chemicals and minimize the negative impact on the environment, as emphasized by the other individual. Through subscriptions to its services, such as through Morshida, where farmers can select from a range of *forfaits* tailored to their needs, the firm generates revenue.

Furthermore, DeepLeaf offers its sophisticated agricultural robot, Agrigo, directly to farmers and agro-food businesses, allowing them to streamline their cultivation management with a single acquisition that encompasses maintenance and cloud-based agreements [29].

In response to these challenges, innovative initiatives are being developed to facilitate sustainable agricultural management in Morocco. It is imperative that local organizations and government agencies collaborate in order to implement environmentally and socially responsible farming practices. The capacity of local farmers to manage natural resources and preserve biodiversity will be enhanced through the implementation of awareness and training programs [30].

Concurrently, it is imperative to endorse sustainable agricultural practices. By endorsing environmentally responsible technologies that facilitate the regeneration of agroecosystems, we contribute to the preservation of biodiversity and guarantee the long-term viability of cultures [31]. Furthermore, the strategy must encompass participatory management and social involvement. The formation of common management standards and agricultural cooperative federations enables communities to collectively manage and safeguard natural resources. The modernization of agriculture has the potential to facilitate sustainable development at the economic and social levels.

By promoting these products on both domestic and foreign markets and ensuring their quality through certification programs, we protect our natural heritage while simultaneously improving the local population's quality of life. The cooperation of the many stakeholders, including local communities and foreign partners, is essential to the success of this integrated approach. Through pooling resources and efforts, we can ensure that agroecosystems continue to thrive and be preserved, preserving their worth for current and future generations.

6.5 CONCLUSION

The application of artificial intelligence (AI) is transforming agricultural productivity and resource management. By analyzing environmental and cultural data, AI enables the optimization of fertilization, irrigation, and disease management. The advent of drones, Internet of Things sensors, and IA-based management systems has enabled farmers to make timely and well-informed decisions.

The implementation of these technologies has the potential to enhance the cultivation of culture while concurrently reducing the excessive utilization of resources and the deleterious effects these have on the environment.

Moreover, in order to enhance the sustainability of agricultural practices, the IA advocates for community engagement and participatory management. Agriculture will ultimately demonstrate enhanced resilience, durability, and productivity in the face of impending challenges due to the integration of these cutting-edge technologies. The modernization of agriculture has the potential to facilitate long-term economic and social progress.

REFERENCES

[1] H. Willer, J. Trávníček, and B. Schlatter, *The World of Organic Agriculture. Statistics and Emerging Trends 2024*. Frick and Bonn: Research Institute of Organic Agriculture FiBL and IFOAM – Organics International, 2024, pp. 1–352. Accessed: Nov. 27, 2024 [Online]. Available: https://orgprints.org/id/eprint/52272/

[2] S. Ghazal, A. Munir, and W. S. Qureshi, "Computer Vision in Smart Agriculture and Precision Farming: Techniques and Applications," *Artificial Intelligence in Agriculture*, vol. 13, pp. 64–83, 2024, https://doi.org/10.1016/j.aiia.2024.06.004.

[3] A. Dahmani and E. E. Akry, "Le plan Maroc vert en faveur des coopératives agricoles, une nouvelle stratégie, une gouvernance renouvelée, de nouveaux moyens," *International Journal of Accounting, Finance, Auditing, Management and Economics*, vol. 3, no. 3–1, Art. no. 3–1, 2022, https://doi.org/10.5281/zenodo.6582405.

[4] R. Balaghi, H. Benaouda, H. Mahyou, and W. Snaibi, "Évaluation de l'impact du changement climatique sur l'agriculture," *African and Mediterranean Agricultural Journal – Al Awamia*, no. 143, Art. no. 143, 2024, https://doi.org/10.34874/IMIST.PRSM/afrimed-i143.48154.

[5] S. Dargaoui, et al., "IoT-Driven Smart Agriculture: Security Issues and Authentication Schemes Classification," in *Proceeding of the International Conference on Connected Objects and Artificial Intelligence (COCIA2024)*, Y. Mejdoub and A. Elamri, Eds., Cham: Springer Nature Switzerland, 2024, pp. 61–66, https://doi.org/10.1007/978-3-031-70411-6_10.

[6] N. Ben-Lhachemi, M. Benchrifa, S. Nasrdine, J. Mabrouki, M. Slaoui, and M. Ade Azrour, "Effect of IoT Integration in Agricultural Greenhouses," in *Technical and Technological Solutions towards a Sustainable Society and Circular Economy*, J. Mabrouki and A. Mourade, Eds., Cham: Springer Nature Switzerland, 2024, pp. 435–445, https://doi.org/10.1007/978-3-031-56292-1_35.

[7] S. Dargaoui, et al., "Internet-of-Things-Enabled Smart Agriculture: Security Enhancement Approaches," in *2024 4th International Conference on Innovative Research in Applied Science, Engineering and Technology (IRASET)*, 2024, pp. 1–5, https://doi.org/10.1109/IRASET60544.2024.10548705.

[8] M. Mohy-Eddine, A. Guezzaz, S. Benkirane, and M. Azrour, "IoT-Enabled Smart Agriculture: Security Issues and Applications," in *Artificial Intelligence and Smart Environment: ICAISE'2022*, Springer, 2023, pp. 566–571.

[9] J. Mabrouki, et al., "Smart System for Monitoring and Controlling of Agricultural Production by the IoT," in *IoT and Smart Devices for Sustainable Environment*, Springer, 2022, pp. 103–115.

[10] M. Azrour, et al., "A Survey of Machine and Deep Learning Applications in the Assessment of Water Quality," in *World Sustainability Series*, vol. Part F2854, 2024, pp. 471–483, https://doi.org/10.1007/978-3-031-56292-1_38.

[11] S. Nasrdine, M. Benchrifa, N. Ben-Lhachemi, J. Mabrouki, M. Slaoui, and M. Azrour, "New Design of an Inclined Solar Distiller for Freshwater Production: Experimental Study," in *World Sustainability Series*, vol. Part F2854, 2024, pp. 447–453, https://doi.org/10.1007/978-3-031-56292-1_36.

[12] W. Shafik and M. Azrour, "Building a Greener World: Harnessing the Power of IoT and Smart Devices for Sustainable Environment," in *Technical and Technological Solutions Towards a Sustainable Society and Circular Economy*, J. Mabrouki and A. Mourade, Eds., Cham: Springer Nature Switzerland, 2024, pp. 35–58, https://doi.org/10.1007/978-3-031-56292-1_3.

[13] C. Hazman, A. Guezzaz, S. Benkirane, and M. Azrour, "Enhanced IDS with Deep Learning for IoT-Based Smart Cities Security," *Tsinghua Science and Technology*, vol. 29, no. 4, pp. 929–947, 2024, https://doi.org/10.26599/TST.2023.9010033.

[14] M. Benchrifa, J. Mabrouki, M. Elouardi, and M. Azrour, "Studying the Effect of Integration Intelligent Dust Detection and Cleaning System on the Efficiency of Monocrystalline Photovoltaic Panels," in *Technical and Technological Solutions Towards a Sustainable Society and Circular Economy*, J. Mabrouki and A. Mourade, Eds., Cham: Springer Nature Switzerland, 2024, pp. 159–169, https://doi.org/10.1007/978-3-031-56292-1_12.

[15] K. El-Moustaqim, J. Mabrouki, M. Azrour, M. Hadine, and D. Hmouni, "Enabling Smart Agriculture Through Integrating the Internet of Things in Microalgae Farming for Sustainability," in *Smart Internet of Things for Environment and Healthcare*, M. Azrour, J. Mabrouki, A. Alabdulatif, A. Guezzaz, and F. Amounas, Eds., Cham: Springer Nature Switzerland, 2024, pp. 209–222, https://doi.org/10.1007/978-3-031-70102-3_15.

[16] H. Hissou, H. Attou, S. Benkirane, A. Guezzaz, M. Azrour, and A. Beni-Hssane, "A Predicted Approach for Solar Radiation Using Multivariate Time Series," in *Advanced Technology for Smart Environment and Energy*, J. Mabrouki and A. Mourade, Eds., Cham: Springer Nature Switzerland, 2024, pp. 269–280, https://doi.org/10.1007/978-3-031-50871-4_18.

[17] S. Li, T. Cui, and W. Viriyasitavat, "Edge Device Fault Probability Based Intelligent Calculations for Fault Probability of Smart Systems," *Tsinghua Science and Technology*, vol. 29, no. 4, pp. 1023–1036, 2024, https://doi.org/10.26599/TST.2023.9010085.

[18] Y. Peng, S. Xu, Q. Chen, W. Huang, and Y. Huang, "A Novel Popularity Extraction Method Applied in Session-Based Recommendation," *Tsinghua Science and Technology*, vol. 29, no. 4, pp. 971–984, 2024, https://doi.org/10.26599/TST.2023.9010061.

[19] N. Meenakshi, et al., "Efficient Communication in Wireless Sensor Networks Using Optimized Energy Efficient Engroove Leach Clustering Protocol," *Tsinghua Science and Technology*, vol. 29, no. 4, pp. 985–1001, 2024, https://doi.org/10.26599/TST.2023.9010056.

[20] P. A. Priya M, P. Karthikeyani, N. Arunfred, M. Hariharan, K. Ramyadevi, and S. Murugan, "IoT and Hydrogen Transport: Revolutionizing Fuel Cell Vehicle Infrastructure," in *2024 4th International Conference on Innovative Practices in Technology and Management (ICIPTM)*, 2024, pp. 1–6, https://doi.org/10.1109/ICIPTM59628.2024.10563939.

[21] K. C. Rath, A. Khang, and D. Roy, "The Role of Internet of Things (IoT) Technology in Industry 4.0 Economy," in *Advanced IoT Technologies and Applications in the Industry 4.0 Digital Economy*, CRC Press, 2024.

[22] M. Emimi, M. Khaleel, and A. Alkrash, "The Current Opportunities and Challenges in Drone Technology," *International Journal of Electrical Engineering and Sustainability*, pp. 74–89, 2023.

[23] Muhammad Naveed Tahir, Yubin Lan, Yali Zhang, Huang Wenjiang, Yingkuan Wang, and Syed Muhammad Zaigham Abbas Naqvi, "Chapter 4 – Application of Unmanned Aerial Vehicles in Precision Agriculture," in *Precision Agriculture*, Q. Zaman, Ed., Academic Press, 2023, pp. 55–70, https://doi.org/10.1016/B978-0-443-18953-1.00001-5.

[24] G. F. Albaaji and V. C. S.S., "Artificial Intelligence SoS Framework for Sustainable Agricultural Production," *Computers and Electronics in Agriculture*, vol. 213, p. 108182, 2023, https://doi.org/10.1016/j.compag.2023.108182.

[25] A. O. Aderibigbe, P. E. Ohenhen, N. K. Nwaobia, J. O. Gidiagba, and E. C. Ani, "Artificial Intelligence in Developing Countries: Bridging the Gap Between Potential and Implementation," *Computer Science & IT Research Journal*, vol. 4, no. 3, Art. no. 3, 2023, https://doi.org/10.51594/csitrj.v4i3.629.

[26] S. Boujdi, A. Ezzahri, M. Bouziani, R. Yaagoubi, and L. Kenny, "A Benchmarking Study of Irrigation Advisory Platforms," *Digital*, vol. 4, no. 2, Art. no. 2, 2024, https://doi.org/10.3390/digital4020021.

[27] R. Ed-Daoudi, A. Alaoui, B. Ettaki, and J. Zerouaoui, "A Machine Learning Approach to Identify Optimal Cultivation Practices for Sustainable Apple Production in Precision Agriculture in Morocco," *E3S Web of Conferences*, vol. 469, p. 00052, 2023, https://doi.org/10.1051/e3sconf/202346900052.

[28] S. Polymeni, S. Plastras, D. N. Skoutas, G. Kormentzas, and C. Skianis, "The Impact of 6G-IoT Technologies on the Development of Agriculture 5.0: A Review," *Electronics*, vol. 12, no. 12, Art. no. 12, 2023, https://doi.org/10.3390/electronics12122651.

[29] V. Kukreja, R. Sharma, S. Vats, and M. Manwal, "DeepLeaf: Revolutionizing Rice Disease Detection and Classification using Convolutional Neural Networks and Random Forest Hybrid Model," in *2023 14th International Conference on Computing*

Communication and Networking Technologies (ICCCNT), 2023, pp. 1–6, https://doi.org/10.1109/ICCCNT56998.2023.10306530.

[30] A. Santoro, et al., "Innovation of Argan (Argania Spinosa (L.) Skeels) Products and Byproducts for Sustainable Development of Rural Communities in Morocco. A Systematic Literature Review," *Biodiversity and Conservation*, 2023, https://doi.org/10.1007/s10531-023-02691-y.

[31] S. Nemade, et al., "Advancements in Agronomic Practices for Sustainable Crop Production: A Review," *International Journal of Plant & Soil Science*, vol. 35, no. 22, Art. no. 22, 2023, https://doi.org/10.9734/ijpss/2023/v35i224178.

7 Utilizing IoT and AI for Soil Health Monitoring and Enhancement in Sustainable Agriculture

Sultan Ahmad, Rupesh Kaushik, Rohit Ghatuary, Amit Kotiyal, Sapna Jarial, and Rajneesh Kumar

7.1 INTRODUCTION

Agriculture, a practice as old as civilization itself, has undergone tremendous advancements over the ages. Traditional agricultural practices have been deeply influenced by the diversity of cultures, climates, and crops across the globe. However, in recent decades, these conventional farming methods have begun to plateau, unable to keep pace with the growing challenges of increasing food demand. To meet the needs of a rapidly expanding global population, there is an urgent need to upgrade these techniques with innovative solutions aimed at increasing yield and productivity. According to the FAO (2017), the world's population is expected to surpass 9 billion by 2050, necessitating a significant increase in food production. However, this comes at a time when arable land is increasingly turning barren due to the overexploitation of natural resources and the excessive use of chemical inputs. Although India's overall agricultural production has increased steadily over the years, the percentage of the population engaged in farming has declined drastically, from 71.9% in 1951 to 45.1% in 2011, and is projected to fall further by 2050 [1]. This declining number of farmers, coupled with the degradation of agricultural land and the need to feed a growing population, has created a critical situation requiring immediate attention. To meet the food demands of 2050, food production must increase by 150% [2]. Adding to these challenges are the profound effects of climate change, which is emerging as a significant abiotic factor negatively impacting agriculture. Unstable and unpredictable weather patterns – including prolonged droughts, frequent floods, and shifting precipitation cycles – are severely reducing crop productivity. These conditions exacerbate issues like soil salinity, nutrient immobilization, and infertility, contributing to widespread land degradation, enhanced desertification, and nutrient-deficient soils. Moreover, the overexploitation of natural resources is rapidly depleting soil health. Intensive farming practices, which depend extensively on chemical fertilizers and pesticides, deplete the soil of vital nutrients, gradually

DOI: 10.1201/9781003527664-7

diminishing its fertility. These challenges underscore the pressing need to shift agriculture toward sustainable, technology-driven solutions. The urgency to address these interconnected issues – rapid population growth, food scarcity, environmental degradation, and soil health deterioration – cannot be overstated. Transitioning the agricultural sector to adopt modern innovations, such as IoT, AI, and precision farming, has become a necessity to achieve sustainable food production and mitigate environmental damage. This strategy is essential for addressing the global challenge of feeding an expanding population while preserving critical natural resources and ensuring ecological harmony.

7.1.1 THE ROLE OF TECHNOLOGY IN SUSTAINABLE AGRICULTURE

With the need of evolution for the advancement of traditional agriculture toward modernization by the integration of technology-driven solutions, artificial intelligence (AI) and the Internet of Things (IoT) are leading this agricultural revolution, providing more sustainable and intelligent crop management techniques. AI and IoT's introduction into agriculture has created new opportunities for sustainability and efficiency. Large volumes of data can be processed and analyzed by AI algorithms, which is essential for crop management decision-making. IoT-based sensors can forecast the best times to plant based on soil and weather trends, improving agricultural production [3, 4]. Farmers now have access to real-time data on various aspects such as crop health, temperature, and soil moisture levels. This allows them to apply pesticides, fertilizers, and water in precise quantities, reducing costs and minimizing waste [5, 6].

The integration of AI and IoT into agriculture significantly enhances sustainability by enabling more efficient resource management, allowing for more efficient fertilizer application and less water use. Sustainable agricultural systems that may satisfy the growing demand for food without destroying natural resources are made possible by AI and IoT [7, 8].

7.2 IMPORTANCE OF SOIL HEALTH IN AGRICULTURE

7.2.1 DEFINITION AND COMPONENTS OF SOIL HEALTH

Soil health is a multifaceted concept that refers to the ability of soil to support plant life, maintain water and air quality, and contribute to overall environmental stability. The ideal conditions for plant development are produced by a confluence of physical, chemical, and biological characteristics that affect soil health. Physical characteristics of soil health include texture, structure, compaction, and porosity influences root penetration, water infiltration, and the soil's overall aeration [9]. Chemically, soil health is distinguished by the availability of vital nutrients such as micronutrients, phosphorus, potassium, and nitrogen, together with the soil's pH and cation-exchange capacity, which control nutrient uptake and balance [10]. Biologically, soil health comprises of activity of bacteria, fungus, earthworms that recycle nutrients, break down organic debris, and support plant health through symbiotic interactions [11].

7.2.2   IMPACT OF SOIL HEALTH ON AGRICULTURAL PRODUCTIVITY

Soil health significantly influences crop yield, total farm production, and agricultural productivity. Crop productivity is strongly impacted by the ideal root development, nutrient absorption, and water retention that healthy soils give. By increasing soil fertility, improving water retention, and lowering insect pressure, research has demonstrated that enhanced soil health achieved through techniques like crop rotation, cover crops, and organic amendments can dramatically boost yields [12]. On the other hand, low soil health which is frequently brought on by intense ploughing, monoculture, and excessive chemical fertilizer use can result in lower crop yields and lower farm profitability. For instance, low water infiltration rates in soils with high degrees of compaction or erosion can result in drought stress and poor crop performance [13]. Moreover, a vicious cycle of decreasing productivity and rising input costs may arise from the long-term consequences of soil degradation, such as decreased microbial diversity and nutrient cycling [14]. Additionally, the quality of the soil is essential for maintaining ecosystem services like carbon sequestration, which lowers atmospheric CO_2 levels and hence slows down climate change. One important advantage that promotes both environmental sustainability and agricultural output is the potential of healthy soils to store organic carbon [15].

7.2.3   CURRENT CHALLENGES IN MAINTAINING SOIL HEALTH

Maintenance of soil health is compromised by several challenges; salinity, nitrogen depletion, and soil erosion are some of the most pressing issues among them. The loss of top soil, which is rich in nutrients and organic matter essential for plant growth, is the result of soil erosion brought on by factors like deforestation, excessive grazing, and poor land management [16]. Prolonged erosion can result in the loss of arable land, worsen water retention capacity, and lower fertility of the soil. Nutrient depletion is another major concern, which occurs when crops are grown intensively without sufficient time for soil nutrient replenishment, reducing the soil's capacity to support robust crops. Additionally, salinity stress is becoming a major threat that significantly impacts crop growth and productivity [17]. Globally, over 20% of irrigated cultivable land is presently affected by excessive salt concentrations, and is steadily growing as a result of intensive chemical fertilizer, pesticide application, excessive irrigation, rising groundwater levels, or inadequate drainage that affects soil health by increasing the concentration of salts in the soil, which inhibits plant growth [18]. Furthermore, maintaining soil health is another shortcoming for traditional soil health monitoring. Long-term soil fertility depends on biological indicators of soil health, including microbial diversity and activity, which are frequently overlooked by current methodologies that mainly concentrate on soil nutrient levels, pH, and texture [19].

7.3   INTEGRATION OF IoT IN SOIL HEALTH MONITORING

The Internet of Things (IoT) is a network made up of physically linked computers equipped with sensors and software that enable them to gather and share data online which makes information sharing easier within a network without requiring direct

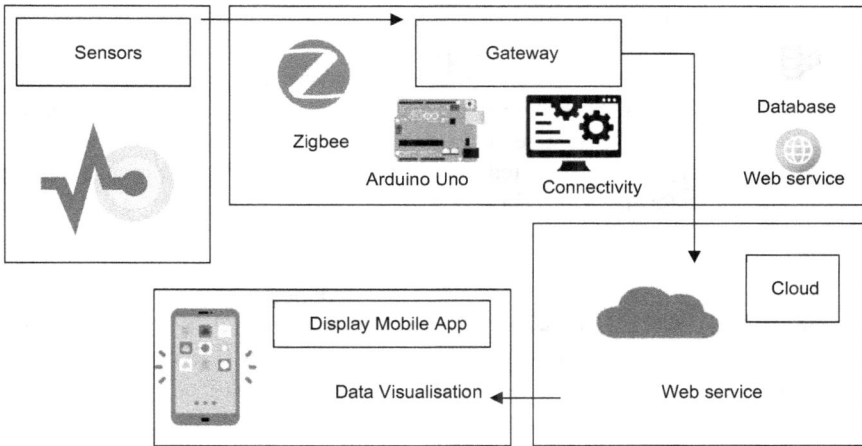

FIGURE 7.1   A diagram of soil health monitoring.

conversation or interaction between humans and computers [20]. IoT technology is transforming agriculture by enabling farmers to monitor and control environmental factors, soil, and crops. Farmers may make quick decisions to optimize agricultural methods, boost production, and conserve resources by using IoT systems to access real-time data. Low-power wide-area networks (LPWAN) or cellular networks are frequently used by these interconnected systems to send data from field-based sensors to central cloud platforms for analysis. IoT has completely transformed the approach to monitoring and managing crops (Figure 7.1). Farmers can now obtain real-time data on a variety of topics, including crop health, temperature, and soil moisture levels, due to a network of sensors placed throughout fields [21]. Precision farming depends on this knowledge as it enables farmers to apply pesticides, fertilizer, and water in the right amounts, cutting expenses and waste. Drones and other IoT devices are crucial for assessing vast farming regions and offering in-depth information on crop health and growth trends [22].

7.3.1   IoT-Based Soil Health Monitoring Systems

IoT combined with smart sensors can transform traditional agricultural methods into smart farming. IoT-based technologies assist in assessing crop quality, soil fertility, fertilizer needs, soil erosion, and soil health benefits from real-time data acquisition and analysis, enabling monitoring of water content, crop quality, nutrient uptake, and site-specific crop production [23]. This approach also helps farmers accurately track the growth and condition of their crops [24]. Additionally, it also helps to evaluate plant illnesses and insect attacks in real time. Farmers and researchers benefit greatly from the real-time data collection and processing provided by IoT-tagged sensors for intelligent crop cultivation, fertilizer application, irrigation, and plant environment control. Water conservation is crucial in the modern agriculture since water is becoming more and more limited. According to the UN Convention to Combat Desertification (UNCCD), nearly 50% of the world's population lives in areas with severe water shortages, and 168 nations will be overtaken by desertification by 2030. The Smart Irrigation Decision Support System (SIDSS), a cutting-edge technique, has demonstrated significant promise for effectively managing the water supply in agricultural areas [25]. Abba et al. [26] evaluated the Smart IoT-based irrigation monitoring and control system's performance and found it to be adequate. Since loamy and sandy soils are the most common varieties that are appropriate for autonomous irrigation farming, they were chosen. The system counter was set between 0 and 300, and various counter values were used to record different moisture measurements. It was observed that the soil tended to absorb water more slowly while irrigation was flowing into it. Curiously, it also held water at one time, which led to the water pump gradually shutting off. This proved that when used to irrigate farms with loamy soil, the intelligent Internet of Things–based irrigation monitoring and control system can function well.

7.3.2 Advantages and Challenges of IoT in Soil Health Monitoring

IoT-based technology has the potential to greatly improve soil health, with successful applications in various small-scale farming areas. However, widespread adoption faces challenges due to the high costs associated with installing IoT sensors and accessories across large agricultural lands. The unclear return on investment and implementation costs add to these challenges. Expenses include purchasing hardware, installing software, maintaining the system, as well as additional costs for energy, system upkeep, service registration, and staffing to manage the integrated hardware and software. In rural farm settings where farmers are less accustomed to modern technology, it may be more difficult to deploy IoT-coupled smart sensors and accessories [27]. Data security and privacy are additional issues that may have a detrimental effect on the widespread deployment of IoT and smart technologies. Cyberattackers could alter data on cloud servers to disrupt automated agricultural systems in farming environments, potentially harming overall farmland productivity and worsening environmental management [28]. A significant challenge with IoT devices is their exposure to harsh outdoor conditions, such as wind, dust, rain, and extreme temperatures, which can lead to unforeseen mechanical failures of these advanced devices.

7.4   ROLE OF AI IN SOIL HEALTH MANAGEMENT

With its cutting-edge methods for tracking, evaluating, and enhancing soil quality, artificial intelligence (AI) is becoming more widely acknowledged for its revolutionary role in soil health management. Large datasets collected from environmental factors, remote sensing technologies, and soil sensors may be processed by AI-driven systems like machine learning algorithms and neural networks to find patterns and forecast trends in soil health [17, 19]. By identifying early indicators of soil deterioration, including erosion, salinity, and nutrient shortages, these systems may suggest customized treatments that maximize crop production and soil fertility. For example, by evaluating soil samples and satellite photos, AI models are used to forecast soil organic carbon sequestration rates, a crucial sign of soil health [15]. The ML and DL structures of AI analyze big data to anticipate and optimize procedures, thereby facilitating information-based decision-making. Furthermore, by combining real-time soil data with IoT devices, AI may improve precision agriculture by facilitating dynamic management techniques like targeted fertilization and irrigation, which reduces environmental impact and increases sustainability [6, 18]. As AI technologies advance, they will provide viable options for managing and monitoring soil health on a wide scale, ultimately assisting international initiatives to stop soil deterioration and guarantee sustainable agricultural output [2, 16].

7.4.1 AI-DRIVEN SOLUTIONS FOR SOIL HEALTH

AI- and IoT-based fertilization techniques assess the moisture content and the spatial distribution of nutrients in the soil. Factors such as crop type, crop requirements, irrigation methods, precipitation, and soil moisture retention all contribute to determining crop water demand. By using wireless sensor systems to monitor air and soil moisture, crop health is improved, and water resources are better managed. Additionally, site-specific nutrient fertilization in smart agriculture precisely calculates nutrient requirements, minimizing the negative environmental and soil effects of overfertilization. Both nutrient deficiencies and excessive fertilizer use can harm soil, plant health, and the broader ecosystem [29]. Variable-rate technologies (VRT) are used in agriculture to forecast the input delivery rates based on a preestablished map derived from GIS, ensuring that inputs are applied in precise quantities, at the right location, and at the right time [30]. Grid soil sampling involves systematically collecting soil samples to create maps for various parameters. These maps are used in VRT to guide a variable-rate applicator, which uses a computer and GPS to adjust fertilizer delivery based on the map's data [31]. The combination of VRT and grid soil sampling can improve soil fertility management and help assess the spatial distribution of nutrients and crop yields [32]. Additionally, IoT-enabled devices like robots, wireless sensors, and drones can precisely detect and manage crop pests through real-time monitoring, modeling, and disease prediction, offering greater efficiency compared to conventional pest control methods [33, 34]. Detection and image processing play a crucial role in IoT-based pest and disease management. During the crop cycle, data on plant health and insect presence are collected from each field using field sensors and remote sensing imagery. IoT-powered automated traps capture, count, and categorize various insect species and then upload the information to the cloud for further analysis [35, 36].

7.4.2 BENEFITS AND LIMITATIONS OF AI IN AGRICULTURE

The integration of AI and IoT technology in agriculture transforms every aspect of traditional farming practices. Smart farming utilizes wireless sensors and IoT to tackle various challenges faced by conventional agriculture, including yield optimization, irrigation, pest control, drought monitoring, and soil fertility loss. However, despite the numerous advantages, obstacles such as limited financial resources, farmers' literacy levels, and system integration issues still impede the widespread adoption of AI and IoT in agriculture. Farmers face financial challenges when they look beyond traditional instruments since the implementation of gadgets and technology is quite expensive. Due to a lack of interest in learning new things or being aware of emerging technologies, farmers in poor countries are typically ignorant and unskilled [37]. For this reason, farmers prefer conventional farming over smart farming. Farmers believe that using a mobile application is too complicated, and they are often unable to identify the symbols since they utilize generic icons based on conventional knowledge. Agri-tech businesses should ensure that farmers can readily comprehend the technology's limitations, but farmers must also be digitally literate to reinforce the benefits of smart farming technologies [38].

7.5   SYNERGISTIC INTEGRATION OF IoT AND AI

7.5.1   How IoT and AI Work Together

The integration of artificial intelligence (AI) and the Internet of Things (IoT) has reached a mature stage, offering immense potential benefits. Often referred to as the driving force behind the Fourth Industrial Revolution, IoT has sparked transformative advancements across diverse fields. Many experts argue that IoT's future inherently depends on AI, predicting that most IoT applications will soon increasingly leverage AI tools and techniques, particularly machine learning and reasoning algorithms [39]. In fact, IoT and AI have already been collaborating across various industries for quite some time. IoT excels in gathering vast amounts of data, while AI processes these data, analyzes them, and makes informed decisions [40]. By recognizing patterns and generating insights, AI enables smarter decision-making. The combination of machine learning and big data analytics has further expanded the possibilities for IoT applications [41].

7.5.1.1   Data Collection via IoT Devices and Analysis through AI Models

Data collection through IoT devices and analysis via AI models plays a pivotal role in the functionality of modern systems like smart grids (SGs) and smart buildings [42]. IoT devices, including sensors and smart meters, gather vast amounts of real-time data, such as building energy demand, interior temperature, and security metrics. This data is then transmitted across networked systems, enabling remote monitoring and configuration to improve energy efficiency, security, and occupant comfort. AI models, particularly those using machine learning techniques, analyze the collected data to forecast energy demand, optimize operations, and make intelligent decisions [43]. By integrating IoT's connectivity with AI's analytical capabilities, smart systems can perform complex tasks autonomously, creating more efficient and sustainable infrastructures – creating a feedback loop for continuous improvement.

7.5.2   Applications in Sustainable Agriculture

AI and IoT have become integral to modern farming, transforming traditional agricultural practices through advanced monitoring and management capabilities.

These technologies are increasingly being applied to ensure high-quality production by enabling real-time monitoring of agricultural ecosystems. However, Smart Sustainable Agriculture (SSA) faces challenges such as the deployment and management of IoT and AI devices, data sharing and governance, system interoperability, and handling large data volumes for analysis and storage.

7.5.3 Contribution to Climate-Smart Agriculture

The Internet of Things (IoT) is a groundbreaking technology that provides efficient and reliable solutions in areas such as smart agriculture and climate change adaptation. By linking billions of smart devices, IoT facilitates the automated monitoring and management of agricultural and environmental sectors, boosting productivity and sustainability [6]. When integrated with artificial intelligence (AI) and blockchain technology, IoT has the capacity to transform agriculture into the "Internet of Smart Agriculture," offering improved control, management, and security within supply chain networks.

This integration facilitates real-time data collection, predictive analysis, and decision-making, enabling Climate-Smart Agriculture (CSA) systems to be more adaptive and efficient. AI and machine learning (ML) optimize resource use, predict climatic impacts, and improve farming systems, while blockchain ensures transparency, traceability, and data integrity in agricultural operations [44].

- *Role in mitigating climate change impacts*
  Machine learning (ML) plays a significant role in mitigating the impacts of climate change by providing robust tools for deriving insights from data and enhancing sustainable energy systems. ML has been applied across diverse domains such as ecosystems, agriculture, buildings and cities, industry, and transportation, addressing critical climate change challenges.

- *Enhancing resource conservation and reducing environmental footprints*
  Advancements in artificial intelligence (AI) and Artificial Intelligence of Things (AIoT) are revolutionizing environmental sustainability by enhancing resource conservation and reducing environmental footprints [39]. These emerging urban paradigms integrate advanced technologies with environmental strategies to address complex ecological challenges. AI and IoT technologies enable smarter eco-cities by optimizing environmental performance, improving resource efficiency, and fostering sustainable urban development.

7.6 SOCIOECONOMIC IMPACTS OF TECHNOLOGICAL INTEGRATION

The integration of IoT and AI in soil health monitoring and enhancement has significant socioeconomic impacts. Positive effects include dematerializing industrial processes and reducing ecological footprints, promoting sustainable resource use and circular economies. However, challenges such as job displacement, privacy concerns, hacking vulnerabilities, rising waste management costs, and diminished human control over cybersystems present complex socioeconomic risks. These technologies also increase information dependency and psychological stress. Addressing these challenges requires adaptive policies and strategies to mitigate their disruptive impacts while maximizing their potential benefits for sustainable agriculture [45].

7.6.1 BENEFITS FOR FARMERS

The integration of IoT and AI into agriculture is revolutionizing farming practices, providing farmers with tools to enhance crop yield and quality for crops. Precision farming, enabled by IoT, gathers critical data on weather, soil quality, and crop growth, automating resource management and optimizing farming techniques [46]. AI analyzes this data to deliver actionable insights, such as ideal planting and harvest times. Disease Detection and Prevention systems leverage AI and machine learning to identify crop diseases early, while Automated Irrigation uses soil moisture sensors to ensure efficient water use. Additionally, IoT and AI improve supply chain management by tracking produce freshness in real time, reducing waste and boosting efficiency [47]. Despite these benefits, challenges such as high initial costs, inadequate Internet connectivity in rural areas, limited digital literacy among farmers, and data privacy concerns remain significant barriers. Overcoming these challenges through investments, supportive policies, and farmer training can unlock the full potential of these technologies, driving productivity and sustainability in agriculture.

7.6.2 CHALLENGES IN ADOPTION

The adoption of IoT and AI technologies in agriculture has the potential to revolutionize farming practices, but several challenges hinder their widespread implementation. While AI tools like precision farming, predictive analytics, automated machinery, smart irrigation systems, crop and soil monitoring, supply chain optimization, weather forecasting, and livestock management can significantly enhance productivity, small-scale and developing farmers often face difficulties in adopting these technologies. The high upfront costs associated with implementing AI technologies make it unaffordable for many farmers [48]. Additionally, these technologies require technical skills, fast Internet connectivity, and costly equipment, which are often lacking in rural areas. As a result, many farmers are unable to access the benefits of AI, leading to wasted resources without tangible outcomes.

7.6.3 POLICY INTERVENTIONS FOR WIDER ADOPTION

Policy intervention plays a crucial role in addressing the socioeconomic impacts of integrating artificial intelligence (AI) and the Internet of Things (IoT) within organizations. Corporate adoption of these technologies often encounters challenges rooted in human social and economic concerns, particularly regarding their effects on individual relevance within a technology-driven framework [49]. To mitigate resistance and promote the successful adoption of AI and IoT, organizations must leverage the principles of sociotechnical systems. Understanding these principles helps align technological convenience with human and organizational needs, thereby enhancing competitiveness and sustainability [50]. The next decade is expected to witness widespread integration of AI and IoT into both private and public sector business processes, driven by their potential to improve effectiveness, competitiveness, sustainability, and efficiency. Policy strategies must focus on adopting technology acceptance and sociotechnical models to navigate socio-psychological and systematic implications.

7.7   FUTURE PROSPECTS AND INNOVATIONS

7.7.1   Emerging Trends in IoT and AI for Agriculture

Smart agriculture integrates advanced information communication and sensing technologies to deliver effective and cost-efficient agricultural services. It leverages a variety of cutting-edge technologies, including wireless sensor networks, Internet of Things (IoT), robotics, agricultural bots, drones, artificial intelligence (AI), and cloud computing [51]. These innovations empower stakeholders in the agricultural sector to make informed managerial decisions, enhancing crop yields and resource efficiency. The transition from traditional to smart agriculture involves distinct deployment architectures and a focus on specific processing stages, supported by various sensors critical to its operation. The integration of these sensing technologies with emerging computational infrastructures further enhances agricultural intelligence. Recent interest in smart agriculture among farmers is fueled by the availability of affordable IoT-based wireless sensors that remotely monitor and report field, climate, and crop conditions [6]. These sensors facilitate efficient resource management by reducing water usage and minimizing toxic pesticide application [52]. The rapid advancements in AI also enable the deployment of autonomous farming machinery and predictive analytics, helping to combat crop diseases and pest infestations more effectively. Collectively, IoT and AI have revolutionized conventional farming, steering it toward sustainability and precision [53]. Despite its potential, smart agriculture faces challenges in widespread adoption, including cost barriers, infrastructure limitations, and the need for technical skills. This field continues to evolve, with ongoing research addressing these hurdles and exploring future trends for global adoption. Efforts focus on making smart agriculture systems more accessible, practical, and scalable to ensure their long-term viability and impact.

- *Integration with robotics and autonomous farming*
  The integration of robotics and autonomous systems (RAS) into farming offers transformative opportunities to enhance agricultural sustainability and efficiency. These technologies enable the optimization of inputs such as fertilizers, seeds, and fuel, reducing environmental impacts on soil and other natural resources [54]. RAS improves the precision and efficiency of agricultural processes and equipment, facilitating better crop care and minimizing farm waste.

- *Use of drones and satellite data for soil health analysis*
  The use of drones and satellite data has emerged as a pivotal tool for soil health analysis and sustainable farming practices. Remote sensing

technologies, including high-resolution satellite sensors and drone-based systems, provide critical diagnostic information about crops and soils [55, 56]. Satellites operating in constellations enable timely and frequent observations with spatial resolutions suitable for small farmlands, aiding regional-scale diagnostics and decision-making [57, 58]. Drones provide an affordable, high-resolution solution for versatile monitoring, collecting data on crop growth, water stress, soil fertility, weeds, diseases, lodging, and 3D topography using optical, thermal, and video imaging. Furthermore, integrating drone-based remote sensing with the precise application of seeds, pesticides, and fertilizers enhances the efficiency of labor and materials, contributing to profitability and sustainable crop production. These advancements underscore the transformative role of smart farming in addressing global food and environmental security challenges [59].

7.7.2 VISION FOR SMART FARMING PRACTICES

- *Adoption of data-driven agriculture globally*
  The global adoption of data-driven agriculture is advancing through the integration of artificial intelligence (AI), despite challenges like low replicability and systematic data collection issues due to field variability [60, 61]. By comparing pilot experiments across diverse fields, weather conditions, and farming techniques, collective knowledge is expanding. Recent research projects in multiple European countries have demonstrated how AI technologies enhance farm-level decision-making by monitoring conditions, optimizing inputs for crops, and reducing water usage and greenhouse gas emissions, all while boosting yields [62]. Although still in its early stages in some regions, these technologies are transforming agricultural practices. Future developments include deploying autonomous and intelligent robots for soil and plant sample collection and improving livestock management, further emphasizing the role of data-driven methods in sustainable agriculture [63].

- *Combining traditional farming methods with advanced technologies*
  The integration of traditional farming methods with advanced technologies is revolutionizing the agricultural industry, making it increasingly data-centric and reliant on precise, advanced technologies like the Internet of Things (IoT). This shift from statistical to quantitative approaches has introduced new opportunities while challenging existing farming practices [64]. The combination of advanced technologies with conventional systems leverages tools such as sensors for land preparation, irrigation, pest, and disease management, and unmanned aerial vehicles (UAVs) for crop monitoring and yield optimization. Comprehensive IoT-based programs enhance every stage of the cropping system, from sowing to harvesting, packaging, and transportation [65]. These technologies are essential for sustainable agriculture, enabling better management practices and maximizing productivity while addressing the challenges of integrating modern tools into traditional systems.

7.8 CONCLUSION

Soil health is crucial for sustainable agriculture as it underpins ecosystem balance, biodiversity, and productivity. It involves maintaining optimal physical, chemical, and biological properties, such as soil texture, organic matter, microbial activity, and nutrient content. Degraded soils, caused by erosion, desertification, and unsustainable practices, threaten food security and environmental health. Restoring soil health through reforestation, organic amendments, and practices like crop rotation and reduced tillage ensures soil resilience, enhances crop productivity, and supports the long-term sustainability of agriculture. The integration of IoT and AI has transformed soil health monitoring and management by enabling precision agriculture practices. Soil sensors, IoT devices, and remote sensing technologies provide real-time data on soil properties and spatial variability. AI-based models and data analytics enhance decision-making by identifying anomalies, optimizing resource allocation, and supporting precision irrigation management. These advancements enable continuous monitoring, timely interventions, and sustainable soil health practices, driving efficiency and productivity in agriculture. The integration of IoT and AI in soil health monitoring underscores the critical need for interdisciplinary research, blending agricultural expertise with technological innovation. Collaborative efforts among policymakers, researchers, and farmers are essential to drive sustainable agricultural practices and ensure food security. A unified call to action is imperative to develop policies, promote knowledge exchange, and implement scalable solutions for resilient farming systems.

REFERENCES

[1] Reddy TK, Dutta M. Impact of agricultural inputs on agricultural GDP in Indian economy. *Theoretical Economics Letters*. 2018;8(10):1840.

[2] Rengasamy P. World salinization with emphasis on Australia. *Journal of Experimental Botany*. 2006;57(5):1017–1023.

[3] Ansar SA, Jaiswal K, Pathak PC, Khan RA. A step towards smart farming: Unified role of AI and IoT. In *International Conference on Computer Vision and Robotics* (pp. 557–578). Singapore: Springer Nature Singapore, 2023.

[4] Heldreth C, Akrong D, Holbrook J, Su NM. What does AI mean for smallholder farmers? A proposal for farmer-centered AI research. *Interactions*. 2021;28(4):56–60.

[5] Khanna A, Kaur S. Evolution of Internet of Things (IoT) and its significant impact in the field of precision agriculture. *Computers and Electronics in Agriculture*. 2019;157:218–231.

[6] Ayaz M, Ammad-Uddin M, Sharif Z, Mansour A, Aggoune EH. Internet-of-Things (IoT)-based smart agriculture: Toward making the fields talk. *IEEE Access*. 2019;7:129551–129583.

[7] Alreshidi E. Smart sustainable agriculture (SSA) solution underpinned by internet of things (IoT) and artificial intelligence (AI). *arXiv preprint arXiv:1906.03106*. 2019.

[8] Tomar P, Kaur G, editors. *Artificial intelligence and IoT-based technologies for sustainable farming and smart agriculture*. IGI Global, 2021. https://doi.org/10.4018/978-1-7998-1722-2

[9] Srinivasan M, Raghavan S, Singh G. Physical properties of soil and their effects on crop yield. *Soil Science and Plant Nutrition*, 2021;67(5):411–421.

[10] Lal R. Enhancing soil health for sustainable agriculture: The role of management practices. *Soil & Tillage Research*. 2020;201:104573.
[11] Usharani KV, Roopashree KM, Naik D. Role of soil physical, chemical and biological properties for soil health improvement and sustainable agriculture. *Journal of Pharmacognosy and Phytochemistry*. 2019;8(5):1256–1267.
[12] Giller KE, Andersson JA, Tittonell P. Soil health and sustainable crop productivity: The role of agricultural practices. *Field Crops Research*. 2021;258:107909.
[13] Nielsen MR, Hansen JL, Skaarup B. Soil compaction and crop yields in intensive agricultural systems. *Agriculture, Ecosystems & Environment*. 2020;296:106913.
[14] Van Der Watt DL, Campbell WL, Cloete D. Effects of soil degradation on productivity and ecosystem services. *Land Degradation & Development*. 2022;33(5):1499–1514.
[15] Smith P, et al. Soil carbon sequestration and its role in sustainable agricultural production. *Nature Sustainability*. 2020;3:354–361.
[16] Lal R, Ahmad M, Singh M. Soil erosion and its impact on agriculture. *Agricultural Systems*. 2020;176:102713.
[17] Gattinger A, Muller A, Pugliese P. Sustainable agricultural practices and their impact on soil health. *Environmental Science & Policy*. 2021;117:52–61.
[18] Qadir M, Oster JD. Crop and irrigation management strategies for saline-sodic soils and waters aimed at environmentally sustainable agriculture. *Science of the Total Environment*. 2004;323(1–3):1–9.
[19] Wang L, Cheng Y, Meftaul IM, Luo F, Kabir MA, Doyle R, Lin Z, Naidu R. Advancing soil health: Challenges and opportunities in integrating digital imaging, spectroscopy, and machine learning for bioindicator analysis. *Analytical Chemistry*. 2024;96(20):8109–8123.
[20] Jebari H, Mechkouri MH, Rekiek S, Reklaoui K. Poultry-edge-AI-IoT system for real-time monitoring and predicting by using artificial intelligence. *International Journal of Interactive Mobile Technologies*. 2023;17(12):149–170.
[21] Abu NS, Bukhari WM, Ong CH, Kassim AM, Izzuddin TA, Sukhaimie MN, Norasikin MA, Rasid AF. Internet of things applications in precision agriculture: A review. *Journal of Robotics and Control (JRC)*. 2022;3(3):338–347.
[22] Ayaz M, Ammad-Uddin M, Sharif Z, Mansour A, Aggoune EH. Internet-of-Things (IoT)-based smart agriculture: Toward making the fields talk. *IEEE Access*. 2019;7:129551–129583.
[23] Sishodia RP, Ray RL, Singh SK. Applications of remote sensing in precision agriculture: A review. *Remote Sensing*. 2020;12(19):3136.
[24] Kaloxylos A, Wolfert J, Verwaart T, Terol CM, Brewster C, Robbemond R, Sundmaker H. The use of future internet technologies in the agriculture and food sectors: Integrating the supply chain. *Procedia Technology*. 2013;8:51–60.
[25] Srilakshmi A, Rakkini J, Sekar KR, Manikandan R. A comparative study on Internet of Things (IoT) and its applications in smart agriculture. *Pharmacognosy Journal*. 2018;10(2).
[26] Abba S, Wadumi Namkusong J, Lee JA, Liz Crespo M. Design and performance evaluation of a low-cost autonomous sensor interface for a smart IoT-based irrigation monitoring and control system. *Sensors*. 2019;19(17):3643.
[27] Ayaz M, Ammad-Uddin M, Sharif Z, Mansour A, Aggoune EH. Internet-of-Things (IoT)-based smart agriculture: Toward making the fields talk. *IEEE Access*. 2019;7:129551–129583.
[28] Neshenko N, Bou-Harb E, Crichigno J, Kaddoum G, Ghani N. Demystifying IoT security: An exhaustive survey on IoT vulnerabilities and a first empirical look on Internet-scale IoT exploitations. *IEEE Communications Surveys & Tutorials*. 2019;21(3):2702–2733.

[29] Köksal Ö, Tekinerdogan B. Architecture design approach for IoT-based farm management information systems. *Precision Agriculture*. 2019;20:926–958.

[30] Batte MT, VanBuren F. Precision farming – Factors influencing profitability. In *Northern Ohio Crops Day Meeting*, Wood County, Ohio, 1999 (Vol. 21).

[31] O'Grady MJ, O'Hare GM. Modelling the smart farm. *Information Processing in Agriculture*. 2017;4(3):179–187.

[32] Fleming KL, Westfall DG, Bausch WC. Evaluating management zone technology and grid soil sampling for variable rate nitrogen application. In *Proceedings of the 5th International Conference on Precision Agriculture* (pp. 16–19). Madison, WI: ASA-CSSA-SSSA, 2000.

[33] Kim S, Lee M, Shin C. IoT-based strawberry disease prediction system for smart farming. *Sensors*. 2018;18(11):4051.

[34] Venkatesan R, Kathrine GJ, Ramalakshmi K. Internet of Things based pest management using natural pesticides for small scale organic gardens. *Journal of Computational and Theoretical Nanoscience*. 2018;15(9–10):2742–2747.

[35] Ennouri K, Kallel A. Remote sensing: An advanced technique for crop condition assessment. *Mathematical Problems in Engineering*. 2019;2019(1):9404565.

[36] Marinelli MV, Scavuzzo M, Giobellina B, Scavuzzo M. Geoscience and remote sensing on horticulture as support for management and planning. *Journal of Agronomy Research*. 2019;2(2):43–54.

[37] Kinyua MW, et al., Plant spatial configurations and their influences on phenological traits of cereal and legume crops under maize-based intercropping systems. *Journal of Sustainable Agriculture and Environment*. 2024;3(2):e212110.

[38] Khan AR, Dubey MK, Bisen PK, Saxena KK. Constraints faced by farmers of Narsing Kheda village of Sihore district. *Young*. 2007;8:16.

[39] Bibri SE, Krogstie J, Kaboli A, Alahi A. Smarter eco-cities and their leading-edge artificial intelligence of things solutions for environmental sustainability: A comprehensive systematic review. *Environmental Science and Ecotechnology*. 2024;19:100330.

[40] Tien JM. Internet of things, real-time decision making, and artificial intelligence. *Annals of Data Science*. 2017;4:149–178.

[41] Paramesha M, Rane NL, Rane J. Big data analytics, artificial intelligence, machine learning, Internet of Things, and blockchain for enhanced business intelligence. *Partners Universal Multidisciplinary Research Journal*. 2024;1(2):110–133.

[42] Koshy S, Rahul S, Sunitha R, Cheriyan EP. Smart grid–based big data analytics using machine learning and artificial intelligence: A survey. *Artificial Intelligence and Internet of Things for Renewable Energy Systems*. 2021;12:241.

[43] Ahmed RA, Hemdan EE, El-Shafai W, Ahmed ZA, El-Rabaie ES, Abd El-Samie FE. Climate-smart agriculture using intelligent techniques, blockchain and Internet of Things: Concepts, challenges, and opportunities. *Transactions on Emerging Telecommunications Technologies*. 2022;33(11):e4607.

[44] Gryshova I, Balian A, Antonik I, Miniailo V, Nehodenko V, Nyzhnychenko Y. Artificial intelligence in climate smart in agricultural: Toward a sustainable farming future. *Access to Science, Business, Innovation in the Digital Economy. ACCESS Press*. 2024;5(1):125–140.

[45] Kasinathan P, Pugazhendhi R, Elavarasan RM, Ramachandaramurthy VK, Ramanathan V, Subramanian S, Kumar S, Nandhagopal K, Raghavan RR, Rangasamy S, Devendiran R. Realization of sustainable development goals with disruptive technologies by integrating industry 5.0, society 5.0, smart cities and villages. *Sustainability*. 2022;14(22):15258.

[46] Bhangar NA, Shahriyar AK. IoT and AI for next-generation farming: Opportunities, challenges, and outlook. *International Journal of Sustainable Infrastructure for Cities and Societies.* 2023;8(2):14–26.

[47] da Costa TP, Gillespie J, Cama-Moncunill X, Ward S, Condell J, Ramanathan R, Murphy F. A systematic review of real-time monitoring technologies and its potential application to reduce food loss and waste: Key elements of food supply chains and IoT technologies. *Sustainability.* 2022;15(1):614.

[48] Qazi S, Khawaja BA, Farooq QU. IoT-equipped and AI-enabled next generation smart agriculture: A critical review, current challenges and future trends. *IEEE Access.* 2022;10:21219–21235.

[49] Sanni AM. *Technological Driven Corporate Social Responsibilities of Telecommunication Industry and Business Communities in Kwara* (Master's thesis, Kwara State University, Nigeria).

[50] Nwankwo W, Adetunji CO, Olayinka AS, Ukhurebor KE, Ukaoha K, Chinecherem U, Chinedu PU, Benson BU. The adoption of AI and IoT technologies: Socio-psychological implications in the production environment. *IUP Journal of Knowledge Management.* 2021;19(1).

[51] Fuentes-Peñailillo F, Gutter K, Vega R, Silva GC. Transformative technologies in digital agriculture: Leveraging Internet of Things, remote sensing, and artificial intelligence for smart crop management. *Journal of Sensor and Actuator Networks.* 2024;13(4):39.

[52] Mutunga T, Sinanovic S, Harrison CS. Integrating wireless remote sensing and sensors for monitoring pesticide pollution in surface and groundwater. *Sensors.* 2024;24(10):3191.

[53] Karunathilake EM, Le AT, Heo S, Chung YS, Mansoor S. The path to smart farming: Innovations and opportunities in precision agriculture. *Agriculture.* 2023;13(8):1593.

[54] Getahun S, Kefale H, Gelaye Y. Application of precision agriculture technologies for sustainable crop production and environmental sustainability: A systematic review. *Scientific World Journal.* 2024;2024(1):2126734.

[55] Rajagopal A, Jha S, Khari M, Ahmad S, Alouffi B, Alharbi A. A novel approach in prediction of crop production using recurrent cuckoo search optimization neural networks. *Applied Sciences.* 2021;11(21):9816.

[56] Ahmad S, Yousuf Uddin M. An intelligent irrigation system and prediction of environmental weather based on nano electronics and Internet of Things devices. *Journal of Nanoelectronics and Optoelectronics.* 2023;18(2):227–236.

[57] Ogungbuyi MG, Mohammed C, Ara I, Fischer AM, Harrison MT. Advancing skyborne technologies and high-resolution satellites for pasture monitoring and improved management: A review. *Remote Sensing.* 2023;15(19):4866.

[58] Linaza MT, Posada J, Bund J, Eisert P, Quartulli M, Döllner J, Pagani A, G. Olaizola I, Barriguinha A, Moysiadis T, Lucat L. Data-driven artificial intelligence applications for sustainable precision agriculture. *Agronomy.* 2021;11(6):1227.

[59] Parra-López C, Abdallah SB, Garcia-Garcia G, Hassoun A, Sánchez-Zamora P, Trollman H, Jagtap S, Carmona-Torres C. Integrating digital technologies in agriculture for climate change adaptation and mitigation: State of the art and future perspectives. *Computers and Electronics in Agriculture.* 2024;226:109412.

[60] Jha S, Routray S, Ahmad S. An expert system-based IoT system for minimisation of air pollution in developing countries. *International Journal of Computer Applications in Technology.* 2022;68(3):277–285.

[61] Kamilaris A, Kartakoullis A, Prenafeta-Boldú FX. A review on the practice of big data analysis in agriculture. *Computers and Electronics in Agriculture.* 2017;143:23–37.

[62] Kumar V, Sharma KV, Kedam N, Patel A, Kate TR, Rathnayake U. A comprehensive review on smart and sustainable agriculture using IoT technologies. *Smart Agricultural Technology*. 2024:100487.

[63] Haque MA, Sonal D, Ahmad S, Kumar K. Enhancing security for internet of things based system. In *International Conference on Communication and Computational Technologies* (pp. 869–878). Singapore: Springer Nature Singapore, 2023.

[64] Azrour M, Mabrouki J, Guezzaz A, Ahmad S, Khan S, Benkirane S, editors. *IoT, Machine Learning and Data Analytics for Smart Healthcare*. CRC Press, 2024.

[65] Azrour M, Dargaoui S, Mabrouki J, Guezzaz A, Benkirane S, Shafik W, Ahmad S. A survey of machine and deep learning applications in the assessment of water quality. In *Technical and Technological Solutions towards a Sustainable Society and Circular Economy* (pp. 471–483). Cham: Springer Nature Switzerland, 2024.

8 Leveraging IoT for Wildlife Deterrence
Smart Solutions for Crop Protection in Modern Farming

Md. Alimul Haque, Deepa Sonal, Sultan Ahmad, and Hikmat A.M. Abdeljaber

8.1 INTRODUCTION

Agriculture is one of the oldest and most vital human activities that lay a foundation for food security, economic development, and societal stability. In an increasingly crowded world, it is important to boost agricultural productivity and sustainability. But the new agricultural sector face many issues that challenge the crop yield and quality – climate change, pests, and wildlife. Another notable but often ignored challenge is crop damage by wild animals. In many parts of the world, wildlife such as deer, wild boar, elephants, and smaller mammals enter farmlands in quest of food sources, thus causing significant losses to farmers and their livelihoods. Traditional approaches in the mitigation of these risks include fencing, chemical deterrents, and physical barriers, which are mostly expensive, labor-intensive, and sometimes damaging to the environment. A new wave of IoT technology that has been brought to the forefront in the past few years provides a new means of agricultural improvement. IoT refers to connecting different devices, such as sensors, cameras, drones, and many other devices, into a network that allows real-time data collection, communication, and response. IoT is becoming an essential part of smart farming practices in agriculture [1–3]. It enables the monitoring of soil conditions, weather patterns, pest activity, and crop health by farmers. However, its potential is much greater than that. It offers a promising solution for managing the threat posed by wildlife with greater precision and lower impact on crops. IoT-enabled devices deployed around crop fields will allow farmers to detect animal activity in real time and deploy targeted responses that do not physically harm the animals or damage the crops. This marks a shift toward smarter, eco-friendly crop protection strategies aligned with the goals of sustainable agriculture. The IoT tools used for crop protection against wild animals basically include sensors that can identify movement or heat signatures, cameras offering live field monitoring, and

DOI: 10.1201/9781003527664-8

deterrent systems such as sound emitters or flashing lights that activate on detection of animal presence. With the use of data analytics and artificial intelligence, these tools detect the presence of animals but also predict the patterns of movement and thus can predict potential threats. For instance, sensors and cameras can be placed along the boundary wall of a farm to detect moving animals and inform the farmer or sends a response through deterrence mechanisms automatically. Solutions above do not come anywhere near the conventional method of pest control because they do not invade the ecosystem; it requires minimal human intervention, and less fencing or chemical deterrents are needed.

This study explores how IoT will provide new means of conserving crops efficiently from wild animals with reduced crop loss. The chapter pays attention to already existing technologies and strategies employed in utilizing IoT for detecting and scaring away the animals. An outline of their efficacy and the problems they experience, coupled with an evaluation of how the technologies may impact the production of sustainable agriculture is provided. These include cases of IoT applications in crop protection – practical implementation and actual successes in a real-world agricultural setting. With the focus on the potential of IoT in this particular area, this chapter hopes to contribute to a greater understanding of how these more advanced technologies can reshape long-practiced agricultural tasks, promoting productivity and responsibility toward the environment. The demand for IoT-based solutions in crop protection is derived from the limitations of current methods and the growing demand for sustainable practices in agriculture. As wild animal populations continue to grow and interact more with human activities, such approaches must be adopted so that these challenges are countered with minimal disruption to the natural ecosystem. Moreover, the scalability as well as the adaptability of IoT proves to be an appropriate technique for large-scale farms besides smallholder farms; indeed, it is very protecting and cost-effective in this regard. The following sections detail the technological, economic, and even ecological aspects of the IoT crop protection system. It demonstrates a holistic view on which these tools are contributing positively to the future of sustainability in agriculture.

8.2 LITERATURE REVIEW

8.2.1 TRADITIONAL CROP PROTECTION TECHNIQUES FROM WILD ANIMALS

Wild animals cause significant damage to agricultural land. They are associated with great economic losses and food insecurity for farmers all over the world. There are four traditional techniques [4] used to protect crops from wildlife – physical barriers, chemical deterrents, biological deterrents, and human intervention. While these techniques have been successful at various levels, each one has some drawbacks and environmental issues of its own.

- *Physical Barriers:* Physical barriers were some of the first used to fence crops from wild animals through walls and trenches. The most widespread form of the physical barrier is fencing, though not possibly applied in situations of recurring intrusion by large animals, including elephants and wild boars.

Electric fencing [5], in particular, is at times imposed more aggressively though endangering both the animals and man. Although relatively efficient, fencing is expensive to install and maintain, particularly for small-scale or low-resource farmers. Moreover, physical barriers can interfere with the normal migration of wild animals, thereby causing an imbalance in the local ecosystems.

- *Chemical Repellents:* There are various chemical repellents that farmers employ to repel animals from their crop fields [6]. These may be sprays, granules, or powders that emit an odor or taste that the animal does not like or detests. Chemical repellents are cheaper, but they pollute the land, water, and food chain with toxic chemicals. The animals might also become resistant to the chemicals after some time, making them ineffective and necessitating stronger chemicals, which may be more toxic. Moreover, chemical applications are labor-intensive and sensitive to time since the chemicals must be reapplied frequently.
- *Biological and Natural Deterrents:* Some farms use natural methods [7], such as growing crops that naturally repel a specific pest or attract beneficial predators. For example, planting marigold flowers around borders can deter deer, while providing habitats for predators within the locality can reduce small pest populations. Biological methods are not effective against pests of larger sizes or those with more versatile food habits. These highly depend on the species to be involved and locale-specific conditions.
- *Human Intervention:* The other classical forms of deterrence, human intervention requires a human farmer or the employed security personnel to move up and down the farm, deterring animals through intimidation. This has been quite effective for small farms but too cumbersome for bigger agricultural sites as they take much time, manpower, and expense. Human intervention has remained very limited in effect because a farm owner will not always cover their area as required. Such an area needs protection constantly against nocturnal and agile animals.

While each of these methods has merits, they often involve trade-offs between cost, labor, environmental sustainability, and effectiveness. Therefore, there is a growing interest in more innovative and scalable solutions, and IoT technology is emerging as an effective and sustainable approach in mitigating these issues.

8.2.2 Overview of IoT Applications in Agriculture

"Smart farming" or "precision agriculture," are commonly used terms that refer to the growth of IoT in agriculture [8], using different tools and techniques to the improve farm efficiency, productivity, and sustainability. The use of IoT applications in agriculture includes the connection of sensors, devices, and data analytic systems for the purpose of providing real-time information as well as an automated response [9–11]. There are several types of IoT tools applied to agriculture such as soil monitoring, pest detection, irrigation management, and recently, crop protection from wild animals [12].

The most frequently applied uses of IoT are in agriculture fields, wherein data is collected on a continuous basis. Sensors can be installed across the field monitoring soil moisture, temperature, humidity, and any form of infestation through pests [13].

Such data are transmitted to a cloud-based server where it will be assessed in real time. The assessments help farmers make decisions for proper irrigation, fertilizers, and pesticides. IoT sensors, which include motion detectors, thermal sensors, and acoustic devices, are helpful in crop protection by indicating the presence of wild animals. These sensors can report animal movement at the periphery of the farm or within specific crop areas. They can send alerts, which can be used for a preemptive response.

- *Automated Deterrents:* This is another emerging technology that combines real-time detection with automated responses. For example, if the sensor detects movement caused by animals, it may then activate lights, sirens, or even ultrasonic sound [14] emitters that can scare the animals away. Such automated systems have reduced the use of physical barriers and human intervention; thus, it is much more efficient and humane compared to others. Some of these mechanisms can be fitted with cameras and deterrent mechanisms and autonomously patrol fields, responding with flight toward the animal as soon as a threat is detected to scare it off. This minimizes the crop damage while saving labor costs and providing coverage where otherwise it is not easily possible.
- *Data Analytics and Predictive Models:* These IoT devices also help the agriculture sector in predictive analytics by collecting large data volumes that can be analyzed to look for patterns. With advancements in artificial intelligence and machine learning [7], IoT systems can detect trends in animal behavior and predict possible threats. Predictive models may be used to inform farmers of peak times for intrusions based on historical data, so that they may strengthen their defenses proactively. The application of IoT in crop protection is quite beneficial since it allows customized and adaptive responses rather than reactive measures.

IoT in agriculture is the integration of tools to enable farmers to manage wild animal threats through a holistic and proactive approach, hence enhancing crop protection strategies with minimal environmental impact and resource usage.

8.2.3   Case Studies of IoT Implementation in Crop Protection

Several case studies are discussed to present the effectiveness and feasibility of IoT-based crop protection solutions against wild animals in various regions. The cases describe the practical applications of IoT tools and demonstrate how the technology is adaptable to different agricultural and environmental settings.

8.2.3.1   Case Study 1: Wildlife Protection System in India

IoT systems have been installed in Indian areas prone to crop damage by elephants [15], to monitor and repel large animals. There are sensors on the fringe of farms that detect elephants through infrared and motion. In case of detecting an elephant, the system comes up with flashing lights and loud alarms that are elephant repellents without harming the elephants. IoT can actually help harmonize agricultural production and also preserve wildlife and hence provide better reduction to crop losses while minimizing human to animal conflicts.

8.2.3.2   Case Study 2: Drones with Thermal Imagery Monitoring Systems in Kenya

Kenya utilizes a drone system fitted with thermal-imaging cameras [13] for prevention of farmland invasion by antelopes. Monkeys are commonly used animals by farmland trespassers to steal feed. Drones with thermal-imaging equipment navigate farms single-handedly scanning to catch heat or warm objects within ranges. Once the potential threat is sensed, the drone descends to the animal and emits a deterrence sound that forces the animal to move away. This will also protect the crops from further damage but, at the same time, avoid unnecessary fencing and, thus retain the landscape of the site and also minimize maintenance charges.

8.2.3.3   Case Study 3: Motion Sensors and Sound Emitters in the United States

IoT-based systems have been installed in farms across the United States to prevent deer from trespassing on farmlands. These systems use motion sensors which are attached to sound emitters [16]. When the deer is detected by the sensors, the emitters will play distress calls or ultrasonic sounds in order to scare the deer away. Due to the aversion that deer have from areas containing such sounds, the losses in crops caused by these animals are kept at bay. In addition, this system can be used with other animal species by adjusting the type and frequency of the sounds and hence forming a versatile method for diverse agricultural settings.

These case studies show the flexibility of IoT applications in crop protection. With real-time data monitoring, automated deterrents, and predictive analytics, IoT provides a robust solution that minimizes the physical and environmental footprint of traditional crop protection methods. As the technology of IoT advances, it is expected to further be applied in agriculture and give even more adaptive and sustainable solutions for protecting crops from wild animals.

8.3   IoT TOOLS AND TECHNOLOGIES IN CROP PROTECTION

IoT technology in farming has transformed the industry by releasing tools that are able to create more accurate crop monitoring and protection. IoT tools and technologies in crop protection have therefore been designed to detect and discourage the threats caused by wildlife without the use of a physical barrier or poisonous chemical. These IoT tools include sensors [17], real-time monitoring devices, automated response mechanisms, and advanced data analytics. These mechanisms ensure crop protection in a manner that minimizes environmental impact. This chapter discusses each of these technologies and tools in detail including their contribution to IoT-based crop protection.

8.3.1   SENSORS AND DETECTORS: MOTION, SOUND, TEMPERATURE, AND INFRARED

Sensors are a major part of IoT-based crop protection systems and detect the movement and acts of animals within agricultural areas. Many types of sensors exist, but each can collect data in a distinct manner, all of which are important to crop protection.

- *Motion Sensors:* It detects all types of motion within an area selected through it, which makes these sensors useful for perimeter coverage of crop fields [18]. Motion detectors send out alerts, either through the light-emitting devices or sound effects if animals enter the zone in the monitored area. One is highly effective for detection of deer and wild boar over open fields; otherwise, an elephant entering the area can be observed due to motion sensors ability to work well under little illumination.
- *Sound Sensors:* Some IoT systems use sound sensors that enable detecting animal calls or footsteps, which indicate that wildlife has entered the area. For example, sounds by elephants and deer can be detected by the presence of sound sensors [19], which act as an early intervention act. Sound sensors are vital in dense vegetation or forest-edge fields where views may not be clear. In addition, integrating sound sensors with algorithms from AI can allow a system to differentiate between sounds generated by various species of animals, thereby allowing for customized responses.
- *Temperature Sensors and Infrared (IR) Sensors:* Temperature and infrared sensors are used to detect heat signatures of animals, proving particularly useful in areas with heavy vegetation or little light. Infrared sensors can sense the specific heat signatures animals release, even when it is dark [14]. IR sensors are also useful in detecting animals that might be hiding behind heavy foliage. Temperature sensors, on the other hand, can provide an indication of the presence of animals through the monitoring of a change in ambient temperature, which is an effective way of scanning the perimeter of fields where animals might find their way.

These sensors provide basic information to be used to monitor real time and activate prompt automatic response mechanisms. Many types of sensors in one IoT system make up an overall or multi-detection mechanism through different agricultural setups and adjustments toward different varieties of threats from wildlife.

8.3.2 REAL-TIME MONITORING IS FACILITATED BY CAMERAS, DRONES, AND TRACKING SYSTEMS

Real-time monitoring of fields is one of the basic advantages that IoT-based crop protection systems offer. A farmer, with this device, always has a view and is able to monitor the fields. For example, cameras, drones, and tracking systems using IoT work in coordination with sensors for live updates or visual confirmation of trespassing animals.

- *Cameras:* General-purpose cameras with night vision capabilities are mounted along field perimeters to monitor unwanted wildlife. Such cameras can record video feeds or images and transmit them to cloud-based platforms or directly to farmers' devices [13, 20]. This gives farmers real-time access to visual information so they can immediately verify any alerts from sensors on the existence of potential threats and determine the severity. Cameras also help track animal behavior and patterns, which can help fine-tune deterrent strategies over time.

- *Drones:* Drones are used more and more with IoT crop protection as they can cover large areas very fast and reach places that are difficult for stationary cameras to view. Equipped with thermal imaging and high-resolution cameras, drones can patrol crop fields, discover animals, and alert farmers [21]. Drones can also act as deterrence mechanisms by moving toward the detected animals and generating sounds or lights to scare those animals away. Their versatility and mobility make drones ideal for large-scale farms and areas with high wildlife activity since they provide both monitoring and active intervention capabilities.
- *IoT Tracking System:* This uses a GPS device with sensors, thus monitoring the animals within and surrounding crop fields. The areas where wildlife tend to encroach on agriculture most are productive areas to apply this technology. By monitoring the system's outputs, it enables intrusion patterns to be known earlier and prepare for intervention much in advance. Connected systems can also communicate directly to government wildlife databases that automatically send alerts when tagged animals are approaching farmland to help reduce human–wildlife conflicts.

Real-time monitoring with such IoT tools makes crop protection strategies more reliable by minimizing the constant need for human intervention and expedites responses to any threats through data-informed decisions.

8.3.3 AUTONOMOUS RESPONSE MECHANISMS: ALARM SYSTEMS, LIGHTING, DRONES

Automated response mechanisms would be an integral part of any IoT crop protection systems as they would enable deterring action to be delivered quickly without human intervention; once a threat is detected through a sensor, the animals can be deterred so that they do not reach or harm the crops.

Sound Deterrents: Sound emitters play distress calls, predator sounds, or ultrasonic noises that are unpleasant to certain animal species. The distress calls and predator sounds frighten away the animals because sensors can detect movement in a specific area. For instance, ultrasonic sounds will deter small mammals, and predator calls will scare deer and wild boar away. They are non-invasive and can target particular animal species; thus, they are friendly to the environment.

Lights: Flashing and strobe lights are also deterrents for nocturnal animals. On detection of an animal, the IoT system flashes its lights, which scares away the animal and makes it leave the area. In general, light deterrents work well on deer and wild boar, who are afraid of bright lights. Light-based deterrents have an added advantage in that they do not disrupt the ecosystem around them and hardly need any maintenance.

Drones as a Deterrent: In addition to being monitoring devices, drones can also be an active form of deterrent by dropping down close to animals that have been detected and making some form of noise or emitting lights [22]. Drones are an effective form of a deterrent because they can approach animals and reach places it would be hard to reinforce with fixed forms of deterrence. These mechanisms can be particularly helpful to the large-scale farms and for those in areas at the risk of wildlife presence because they can minimize physical barriers and reduce human–wildlife conflict, ensuring greater coexistence.

IoT system reduces the need for response mechanisms and thus limits their physical barriers, promoting the harmonious coexistence by minimizing human–wildlife conflict.

8.4 METHODOLOGY

In this research, a comparative experimental design is used. Two crop fields with equal environmental conditions and wildlife are chosen for the study; one is equipped with IoT-based protection systems (Field A), and the other is not (Field B). Observations will be collected from these two fields. By comparing these fields, one can estimate the effects of IoT tools in reducing the damage to crops caused by wild animals and the crop health variations, damage percentage, and wildlife activity that occurs in both setups.

8.4.1 IDENTIFICATION AND SITE PREPARATION

Two fields with the same crop type are used for this test, situated in a close geographic region to have almost similar environmental conditions. Field A shall be provided with an integrated IoT-based crop protection system. Field B will be dependent on conventional or no deterrents for crop protection. Fields selected must have a past record of damage caused to crops by wild animals so that a strong basis of observation for effectiveness of an IoT system may be provided.

8.4.1.1 Field A (IoT-Enabled)

IoT-enabled field will comprise of the following:

- *Sensors:* Motion, infrared (IR), and temperature sensors are installed at strategic points along the periphery of the field to track animal movement.
- *Real-Time Monitoring:* Cameras and drones are used for real-time monitoring of animal activity.
- *Automated Deterrents:* Sound and light emitters and drones are set up to activate automatically when an intrusion is detected.
- *Data Analytics Platform:* AI-driven software that analyzes data and adjusts deterrents based on patterns and animal behavior.

8.4.1.2 Field B (Without IoT)

This will be the control and have no IoT-based systems. The activity of all animals and crop damage will be recorded manually. Farmers may use their traditional deterrent techniques such as fencing or scarecrows, if required.

8.4.2 Data Collection

The data collection shall be done through the measurement of animal activity, crop damage, and success rates of intervention in both the fields over a period of 6 months, thus including every season and all levels of wildlife activity.

8.4.2.1 For Field A (With IoT)

- *Infiltration Data:* Each infiltration will be captured by IoT sensors and cameras that record time, duration of the infiltration, and type of animal.
- *Response Using Deterrent:* Response involving the automated deterrent, for example, if it emits a sound, light, or drone, will be measured by recording the type and whether the animal leaves the area.
- *Crop Damage:* After recording the incident of infiltration, damage caused to crops resulting from the infiltration in the affected area or yield loss for crop is estimated.

Weather and Environmental Data: Through the IoT sensors, whether the natural environment has a bearing on the presence or activities of animals is determined.

8.4.2.2 For Field B (Without IoT)

- *Data on Intrusions:* By visual inspection of the field or signs such as tracks or droppings, data regarding intruding animals will be gathered manually.
- *Traditional Deterrents:* Use of traditional deterrents, like scarecrows or fencing, and how these help in deterring animal intrusions.
- *Crop Damage:* Crop damage will be measured after each recorded entry, comparing the effect of crop health or yield.

Data collected on both fields will be uploaded onto a common central data bank to make easy comparisons between the two fields that either have or do not have IoT protection.

Table 8.1 aims to show significant reductions of wildlife intrusions and crop damage in Field A relative to Field B, where the traditional method or no deterrent is implemented. Upon a positive conclusion, the effectiveness of this IoT technology shall be given credence to its utility in developing sustainable and non-intrusive crop protection.

Table 8.2 clearly demonstrates that IoT-based systems offer substantial improvements in terms of intrusion reduction, response time, crop damage prevention, and operational efficiency, resulting in higher farmer satisfaction and potential economic benefits.

Table 8.2 illustrates a marked improvement in crop protection efficiency when IoT devices are utilized. Fields with IoT systems experienced an 86% reduction in total crop damage and achieved intrusion deterrence 90% faster than traditional methods, showcasing the proactive response capability of IoT technology. Additionally, IoT-equipped fields required significantly less labor, with a 67% reduction in weekly monitoring time, allowing for streamlined operations and cost-savings. Farmer

TABLE 8.1

Data Showing Responses from Both Types of Fields

Observation ID	Date	Field	Intrusion Type	Animal Species	Number of Intrusions	Detected by	Response Type (Field A)	Response Success Rate (Field A)	Crop Damage (m²)	Crop Damage (%)
1	2024-05-01	Field A (with IoT)	Attempted	Deer	2	Motion sensor	Sound, light	90%	0	0%
2	2024-05-01	Field B (without IoT)	Successful	Wild boar	1	Visual observation	N/A	N/A	12	5%
3	2024-05-05	Field A (with IoT)	Attempted	Elephants	3	Camera	Sound, drone	100%	0	0%
4	2024-05-05	Field B (without IoT)	Successful	Elephants	2	Visual observation	N/A	N/A	30	12%
5	2024-05-10	Field A (with IoT)	Attempted	Small mammals	5	IR sensor	Light	80%	2	1%
6	2024-05-10	Field B (without IoT)	Successful	Small mammals	4	Visual observation	N/A	N/A	6	3%
7	2024-05-15	Field A (with IoT)	Attempted	Wild boar	1	Drone	Sound	85%	0	0%
8	2024-05-15	Field B (without IoT)	Successful	Wild boar	1	Visual observation	N/A	N/A	15	6%
9	2024-05-20	Field A (with IoT)	Attempted	Deer	2	IR sensor	Light, drone	95%	1	0.5%
10	2024-05-20	Field B (without IoT)	Successful	Deer	2	Visual observation	N/A	N/A	10	4%

TABLE 8.2
Data Showing the Increase in Efficiency of Farming Techniques with IoT Devices

Metric	With IoT Devices (Field A)	Without IoT Devices (Field B)	Efficiency Improvement (%)
Intrusion success rate	8%	68%	60%
Average response time (seconds)	2–3 seconds	10–15 minutes (manual)	~90% faster
Average crop damage per intrusion (% of area)	1%	6%	83% reduction
Total crop damage over study period	5% of crop area	35% of crop area	86% reduction
Farmer satisfaction rating	4.5 out of 5	2.5 out of 5	80% increase
Deterrent activation success rate	92%	N/A (manual deterrents)	N/A
Labor requirement (hours/ week)	5 hours (monitoring and maintenance)	15 hours (manual patrolling)	67% reduction
Return on investment (ROI)	High (due to reduced crop loss)	Moderate	Significant improvement

satisfaction also rose by 80% in IoT fields, reflecting increased confidence in the technology's effectiveness and the potential for a high return on investment by preserving crop yield. These results indicate that IoT adoption substantially enhances productivity and economic viability in crop protection.

8.5   RESULTS AND DISCUSSION

This fair comparison of Field A, IoT-activated crop protection, and Field B, without IoT, brings forth the power of IoT concerning minimal crop damage caused by wildlife. For instance, in Field A, where IoT-activated deterrents such as sound emitters, lights, drones, and infrared sensors were deployed to ward off wildlife, an intrusion of only 8% through 50 attempts was successful. In sharp contrast, 68% prevailed over Field B with conventional means of deterrent. The damage on crops was also significantly lower in Field A at a mean rate of 1% per invasion and only amounted to 5% over the study period compared to that of Field B at 35%. The response from the IoT system was very fast, taking less than 2–3 seconds, though the setup costs at the onset were high, whose ability to prevent losses was postulated to ensure long-term profitability [23, 24]. Very high satisfaction among farmers was also indicated in the surveys conducted in Field A, rating 4.5/5, as a sign of confidence that IoT is both efficient and scalable for sustained crop protection. This study has brought out some significant advantages of crop protection against agricultural pests using the tools of IoT. Traditional methods, like fencing and scarecrows, realize little with continuous maintenance, whereas IoT-based systems provide proactive automated monitoring coupled with effective deterrent responses. IoT sensors, for example, motion sensors and infrared sensors, could sense animals in real time and activate applicable deterrents as dependent upon the particular threat and optimized through AI-based data processing that learns from previous animal behaviors. This adaptability is particularly valuable for large fields where manual monitoring is impractical. The use of IoT also reduces crop damage and economic loss, which is represented by relatively lower damage rates in the fields applying IoT [25–27]. In addition, it promotes sustainable agriculture due to reduced use of destructive chemical repellers. Though high initial cost, some operational challenges such as impacts of weather or sensor calibration faults exist, IoT-based solutions are cost-effective, and future potential for improvement is possible with advanced AI, such that predictions are possible through environment data – temperature and humidity.

8.6   CONCLUSION

This study thus underlines the obvious role that IoT can play in crop protection against wildlife while offering a sustainable alternative to traditional means of scaring wildlife away. A comparative analysis of the application of such systems, as set up by the use of IoT, indicates a significant reduction in the intrusion of wildlife and crop damage. The adaptive proactive response offered in real time through monitoring and the use of automated deterrents and data analytics have also been shown. The main benefits will be increased detection and response, considerable reduction in crop damage, and economic benefits because of yield preservation – all of which make the adoption of IoT especially worthwhile for bigger farms and areas under frequent wildlife threat [17, 28]. However, to overcome the barriers of high initial costs, technical complexities, and environmental, the prospects of using IoT must be broadened across a variety of agricultural settings [23–25]. Ultimately, the technological advancement of the field into IoT technology seems promising as a solution to issues of crop protection in wildlife-related issues and advancement into sustainable agriculture by reducing dependence on noxious forms of deterrence and increasing yields [26–27]. Future development in AI and IoT integration will further refine such systems into establishing IoT as a crucial element of precision agriculture.

REFERENCES

[1] S. Dargaoui et al., "IoT-Driven Smart Agriculture: Security Issues and Authentication Schemes Classification," in *Proceeding of the International Conference on Connected Objects and Artificial Intelligence (COCIA2024)*, Y. Mejdoub and A. Elamri, Eds., Cham: Springer Nature Switzerland, 2024, pp. 61–66. https://doi.org/10.1007/978-3-031-70411-6_10.

[2] K. El-Moustaqim, J. Mabrouki, M. Azrour, M. Hadine, and D. Hmouni, "Enabling Smart Agriculture through Integrating the Internet of Things in Microalgae Farming for Sustainability," in *Smart Internet of Things for Environment and Healthcare*, M. Azrour, J. Mabrouki, A. Alabdulatif, A. Guezzaz, and F. Amounas, Eds., Cham: Springer Nature Switzerland, 2024, pp. 209–222. https://doi.org/10.1007/978-3-031-70102-3_15.

[3] S. Dargaoui et al., "Internet-of-Things-Enabled Smart Agriculture: Security Enhancement Approaches," in *2024 4th International Conference on Innovative*

Research in Applied Science, Engineering and Technology (IRASET), May 2024, pp. 1–5. https://doi.org/10.1109/IRASET60544.2024.10548705.

[4] K. Mishra, M. K. Mishra, S. K. Shrivastava, B. K. Mishra, and D. Sonal, "An Invisible Fence: Laser Fencing System for Protecting Crops," pp. 1–8, 2022, https://doi.org/10.17605/OSF.IO/Z3BTC.

[5] N. Srikanth et al., "Smart Crop Protection System from Animals and Fire using Arduino," *International Journal of Engineering Research in Electronics and Communication Engineering*, vol. 6, no. 4, pp. 17–21, 2019.

[6] D. Sonal, K. Mishra, M. K. Mishra, S. K. Shrivastava, and B. K. Mishra, "Analysis of Impact of Repelling Sound on Animals," *NeuroQuantology*, vol. 20, no. 12, p. 1335, 2022.

[7] D. Sonal et al., "Agri-IoT Techniques for Repelling Animals from Cropland," p. 12681, 2022. https://doi.org/10.3390/MOL2NET-08–12681.

[8] S. Giordano, I. Seitanidis, M. Ojo, D. Adami, and F. Vignoli, "IoT Solutions for Crop Protection against Wild Animal Attacks," in *2018 IEEE International Conference on Environmental Engineering, EE 2018 – Proceedings*, vol. 1, no. 710583, 2018, pp. 1–5. https://doi.org/10.1109/EE1.2018.8385275.

[9] N. Ben-Lhachemi, M. Benchrifa, S. Nasrdine, J. Mabrouki, M. Slaoui, and M. ade Azrour, "Effect of IoT Integration in Agricultural Greenhouses," in *Technical and Technological Solutions Towards a Sustainable Society and Circular Economy*, J. Mabrouki and A. Mourade, Eds., Cham: Springer Nature Switzerland, 2024, pp. 435–445. https://doi.org/10.1007/978-3-031-56292-1_35.

[10] M. Mohy-eddine, A. Guezzaz, S. Benkirane, and M. Azrour, "IoT-Enabled Smart Agriculture: Security Issues and Applications," in *Artificial Intelligence and Smart Environment: ICAISE'2022*, Springer, 2023, pp. 566–571.

[11] J. Mabrouki et al., "Smart System for Monitoring and Controlling of Agricultural Production by the IoT," in *IoT and Smart Devices for Sustainable Environment*, Springer, 2022, pp. 103–115.

[12] D. R. Reddy, M. Kavya, S. Dharani, S. S. Tumpudi, P. Kodali, and N. Sandhya, "Design and Development of a Low-Cost Crop Protection System Using the Internet of Things and Machine Learning," in *2022 IEEE International Symposium on Smart Electronic Systems (iSES)*, Dec. 2022, pp. 610–614. https://doi.org/10.1109/ISES54909.2022.00134.

[13] K. Balakrishna, F. Mohammed, C. R. Ullas, C. M. Hema, and S. K. Sonakshi, "Application of IOT and Machine Learning in Crop Protection against Animal Intrusion," *Global Transitions Proceedings*, vol. 2, no. 2, pp. 169–174, 2021. https://doi.org/10.1016/j.gltp.2021.08.061.

[14] M. Kumar Mishra and D. Sonal, "Object Detection: A Comparative Study to Find Suitable Sensor in Smart Farming," *Springer Proceedings in Complexity*, pp. 685–693, 2022. https://doi.org/10.1007/978-3-030-99792-2_58/COVER.

[15] M. G. P. M. Samarasinghe, "Use of IOT for Smart Security Management in Agriculture," vol. 9, no. 978, pp. 65–73, 2019.

[16] P. Nair, K. Nithiyananthan, and P. Dhinakar, "Design and Development of Variable Frequency Ultrasonic Pest Repeller," *Journal of Advanced Research in Dynamical and Control Systems*, vol. 9, no. 12, pp. 22–34, 2017.

[17] Y. S. Suh, "Laser Sensors for Displacement, Distance and Position," *Sensors (Switzerland)*, vol. 19, no. 8, 2019. https://doi.org/10.3390/s19081924.

[18] J. Bzai et al., "Machine Learning-Enabled Internet of Things (IoT): Data, Applications, and Industry Perspective," 2022. https://doi.org/10.3390/electronics11172676.

[19] D. Sonal, S. Haque, M. M. Nezami, and A. Balqarn, "An IoT-Based Model for Defending against the Novel Coronavirus (COVID-19) Outbreak," *Solid State Technology*, vol. 63, no. 2, 2020.

[20] M. Zhihong, M. Yuhan, G. Liang, and L. Chengliang, "Smartphone-Based Visual Measurement and Portable Instrumentation for Crop Seed Phenotyping," *IFAC-PapersOnLine*, vol. 49, no. 16, pp. 259–264, 2016. https://doi.org/10.1016/j.ifacol.2016.10.048.

[21] M. Jagadeesan, P. A. Selvaraj, M. Kumar, V. Kumar, and G. Kumar, "IOT Enabled for Smart Farming," *New Frontiers in Communication and Intelligent Systems*, pp. 425–435, 2021. https://doi.org/10.52458/978-81-95502-00-4-45.

[22] M. A. Haque, D. Sonal, S. Haque, and K. Kumar, "Internet of Things for Smart Farming," 2021.

[23] M. Kumar Mishra and D. Sonal, "Object Detection: A Comparative Study to Find Suitable Sensor in Smart Farming," pp. 685–693, 2022. https://doi.org/10.1007/978-3-030-99792-2_58.

[24] M. A. Haque, S. Haque, D. Sonal, K. Kumar, and E. Shakeb, "Security Enhancement for IoT Enabled Agriculture," *Materials Today: Proceedings*, 2021. https://doi.org/10.1016/J.MATPR.2020.12.452.

[25] Md. A. Haque, D. Sonal, S. Ahmad, and K. Kumar, "Enhancing Security for Internet of Things Based System," 2023, pp. 869–878. https://doi.org/10.1007/978-981-99-3485-0_68.

[26] K. M. Mohit, K. M. Shailesh, K. S. Binay, K. Mishra, and D. Sonal, "Analysis of Impact of Repelling Sound on Animals," *An Interdisciplinary Journal of Neuroscience and Quantum Physics*, vol. 20, no. 12, pp. 1335–1341, 2022. https://doi.org/10.14704/NQ.2022.20.12.NQ77111.

[27] D. Sonal, K. Mishra, A. Haque, and F. Uddin, "A Practical Approach to Increase Crop Production Using Wireless Sensor Technology," *LatIA*, vol. 2, pp. 10–10, 2024. https://doi.org/10.62486/LATIA202410.

[28] K. Mishra, M. K. Mishra, S. K. Shrivastava, B. K. Mishra, and D. Sonal, "An Invisible Fence: Laser Fencing System for Protecting Crops," *Jilin Daxue Xuebao (Gongxueban)/Journal of Jilin University (Engineering and Technology Edition)*, vol. 41, 2022.

9 Advancing Water Quality Assessment through Artificial Intelligence Techniques

Anas Kabbori, Azidine Guezzaz,
Said Benkirane, and Mourade Azrour

9.1 INTRODUCTION

Water is a fundamental element of our planet and represents the very essence of life itself. Its presence is ubiquitous, found in vast oceans, flowing rivers, and even within the smallest cellular structures of living organisms. Water goes beyond merely quenching our thirst; it plays a vital role in numerous biological processes. As a solvent, it facilitates the chemical reactions necessary for metabolism, allowing organisms to extract energy from food. Water is also essential for regulating temperature, supporting cellular function, and transporting nutrients and waste within living systems. In plants, it is crucial for photosynthesis, making it possible for them to convert sunlight into energy and thus sustain not only themselves but also the entire food web.

Beyond its biological importance, water serves as a fundamental catalyst for ecosystems, driving nutrient cycles, supporting biodiversity, and shaping the landscapes we inhabit. It plays a critical role in maintaining habitats, such as wetlands and rivers, which provide refuge for countless species. Water's influence extends into every facet of human existence as well. It is integral to agricultural productivity, where irrigation practices rely on water to ensure crop yields and food security. In industry, water is indispensable for processes ranging from cooling machinery to manufacturing goods. Moreover, it is crucial for energy generation, particularly in hydroelectric power, which harnesses moving water to produce electricity. Given its importance in both natural and human systems, understanding and protecting water is essential for sustaining life and ensuring a healthy planet for future generations.

However, water holds the key to public health and societal well-being, as access to clean and potable water is a prerequisite for combating disease, ensuring sanitation, and fostering equitable development. The landscape of AI techniques has evolved rapidly, offering a diverse set of tools for addressing the complexities inherent in water quality monitoring. Machine learning algorithms, including supervised and unsupervised learning, deep learning, and ensemble methods, have gained

 DOI: 10.1201/9781003527664-9

prominence. Manisha Koranga et al. [1] provided an insightful review of AI techniques in water quality assessment, advocating for the development of a predictive model using machine learning algorithms to forecast the water quality index and classify water quality. The study focuses on four key water parameters: temperature, pH, turbidity, and coliforms. Notably, the application of multiple regression algorithms proves effective in predicting the water quality index, while the adoption of artificial neural networks emerges as a highly efficient method for classifying water quality.

9.2   APPLICATIONS IN WATER QUALITY PREDICTION

Water quality forecasting applications are essential for the efficient monitoring and management of water resources. They use advanced technologies and predictive models to evaluate water quality in a wide range of environments, from rivers and streams to lakes and freshwater sources. Real-time collectors are capable of recording crucial parameters rapidly, such as pH, turbidity, and dissolved oxygen, enabling the detection of pollutants and changes in water quality. In addition, such systems contribute to environmental protection, public health, agricultural management, and urban planning by predicting potential pollution or contamination trends. Eventually, water quality forecasting plays a vital role in guaranteeing safe and sustainable water resources for ecosystems and human populations.

The incorporation of artificial intelligence (AI) in predicting water quality parameters has garnered significant attention in recent research efforts. Notably, a study by Ahmed et al. [2] focuses on supervised machine learning algorithms for estimating the Water Quality Index (WQI) and determining the Water Quality Class (WQC). This methodology employs temperature, turbidity, pH, and total dissolved solids as input parameters, with gradient boosting and polynomial regression yielding efficient WQI predictions. Rana et al. [3] utilize a dataset incorporating critical constraints like temperature, pH, dissolved oxygen (DO), conductivity, total dissolved solids (TDS), turbidity, and chlorides (Cl$^-$) to categorize and predict WQI. This investigation employs artificial neural networks (ANN) and long short-term memory (LSTM) models, assessing their performance using metrics, including mean absolute error (MAE), mean squared error (MSE), and coefficient of determination (R^2). Heat maps and correlation graphs are introduced to visually illustrate connections between water quality measures, employing color-coded values to represent different water quality levels. These studies collectively underscore the efficacy of AI and machine learning models in advancing water quality prediction methodologies, offering valuable insights for sustainable water resource management and environmental conservation; another study, exemplified by authors like Lingxuan Chen et al. [4], tackles water pollution issues in rivers through a novel Adaptive Evolutionary Artificial Bee Colony-Back Propagation Neural Network (AEABC-BPNN) model. This model, based on machine learning and intelligent optimization algorithms, exhibits robustness and enhanced convergence speed, surpassing traditional models. A comprehensive review by Yan et al. [5] provides a thorough analysis of over 170 studies within the last five years, categorizing ML-based predictions into indicator and water quality index segments.

The review highlights advancements in methodologies for acquiring water quality data and explores research trends in machine learning algorithms, emphasizing areas such as hydrodynamic water quality coupling and effective data processing. It acknowledges challenges in generalizing results for coastal water quality prediction and recommends diversifying data sources for increased accuracy. In the realm of neural network modeling, Kulisz et al. [6] demonstrate the efficacy of predicting the Surface Water Quality Index using physicochemical parameters. Their study showcases the flexibility of artificial neural network (ANN) models, providing accurate predictions with reduced physicochemical parameters, making them practical in resource-intensive scenarios. In a unique approach, Al-Janabi et al. [7] propose the Iraqi Water Quality Index (IQWQI) tailored to Iraqi aquatic systems. The study incorporates a Delphi method to determine optimal parameters and their weights, introducing an Environmental Risk Index (ERI) to address index eclipse issues, showcasing the model's efficiency in assessing water quality dynamics. In addressing challenges associated with existing water quality index calculations, Mamat et al. [8] propose an optimized support vector machine (SVM) methodology. The model exhibits exceptional replication of the Department of Environment (DOE) Malaysia's water quality index, showcasing efficiency in terms of time and cost, especially in resource-constrained environments. Subsequently, Hoque et al. [9] explore the application of eight machine learning regression algorithms for predicting water quality index in Indian rivers, emphasizing the significance of regression algorithms and specific feature sets in outperforming existing models. Additionally, See Leng Chia et al. [10] introduce a novel approach by integrating the least-square support vector machine (LSSVM) with advanced optimization algorithms. The study demonstrates the superiority of the self-adapting parameter adjustment and mix mutation strategy (SMWOA) LSSVM model in predicting the water quality index, highlighting the potential of advanced optimization algorithms. On Bhavani River, Nair et al. [11] utilized machine learning algorithms for predictive modeling of the water quality index. Their results reveal the efficacy of MLP regressor and MLP classifier in WQI prediction, outperforming other models and suggesting potential applications in hybrid models. In a study by Mbachu et al. [12], principal component analysis is employed to develop water quality models for selected ponds in the Aboh-Mbaise local government area of Imo state. The multilinear regression model demonstrates high accuracy, providing a significant tool for predicting the Water Quality Index (WQI). Focusing on groundwater quality in the Vellore district, Vijay et al. [13] leverage artificial neural networks (ANN) to predict the WQI. Their study incorporates three activation functions and identifies influential variables through sensitivity analysis, contributing insights for future work. Moving to sub-watersheds in Iran, Singh et al. [14] employ artificial neural networks and generalized neural network (GRNN) to predict the water quality index (WQI). Their results emphasize the efficacy of soft computing techniques, particularly ANN, in predicting WQI with high accuracy, offering an efficient alternative to chemical analysis. Addressing the challenges of assessing groundwater quality in arid regions, Derdour et al. [15] employ support vector machine (SVM) and K-nearest neighbors (KNN) classifiers. Their findings underscore the potential of AI, specifically SVM, as an effective tool for evaluating water quality in arid zones, providing valuable insights for decision-makers in groundwater management. In a different context, Volf et al. [16] focus on enhancing the treatment processes of a drinking water treatment plant (DWTP) in Croatia through the prediction of the water quality index. Their study employs various parameters and models predicting WQI with different time steps, highlighting the importance of predictive models in managing treatment processes during periods of high water demand. These studies collectively contribute to the growing body of knowledge on AI applications in water quality prediction, offering insights into various methodologies, models, and considerations for effective water resource management and sustainable development.

9.3 ADVANTAGES OF USAGE OF AI IN WATER QUALITY PREDICTION

9.3.1 ENHANCED ACCURACY AND EFFICIENCY

AI algorithms can analyze vast amounts of data rapidly and accurately, allowing for precise predictions of water quality changes. These algorithms utilize advanced techniques, such as machine learning and deep learning, to identify complex patterns and correlations within the data that may not be apparent through conventional analytical methods [17]. As a result, they can process inputs from various sources, including real-time sensor readings, historical datasets, and environmental variables, to generate insights about current and future water quality conditions.

This efficiency not only enhances the accuracy of predictions but also significantly reduces the time needed to detect and respond to potential contamination or pollution events compared to traditional methods. In scenarios where every second counts – such as during harmful algal blooms or chemical spills – AI-driven systems can provide immediate alerts to relevant authorities, allowing for timely interventions that could mitigate environmental impact and protect public health [18].

Moreover, the speed and precision of AI predictions facilitate proactive management strategies, enabling stakeholders to implement preventive measures rather than merely reacting to water quality issues after they arise. For example, with accurate forecasts of potential pollutant influx, water resource managers can adjust treatment processes or initiate targeted assessments to prevent harmful substances from entering the water supply [19].

Ultimately, the integration of AI into water quality monitoring not only improves operational efficiency but also enhances overall resource management, ensuring a safer and more sustainable approach to maintaining the health of aquatic ecosystems and community water sources.

9.3.2 REAL-TIME MONITORING

AI-powered systems can process data from sensors and monitoring devices in real time, providing immediate insights into water quality parameters such as pH, turbidity, and dissolved oxygen. By continuously analyzing this data, these systems can detect fluctuations and anomalies that may indicate potential issues, such as contamination or nutrient overload. This capability allows for timely interventions to protect public health and environmental ecosystems, as stakeholders can respond quickly to emerging threats [20–22].

For instance, if a sudden spike in turbidity levels is detected, indicating possible sediment runoff or pollution, the system can alert water quality managers to investigate the source and take corrective actions [23–25]. Similarly, real-time monitoring of dissolved oxygen levels can help identify conditions that may lead to hypoxia, which can be detrimental to aquatic life. By providing these immediate insights, AI systems enable proactive management strategies that can prevent environmental degradation and safeguard aquatic habitats [26–28].

Furthermore, the integration of AI with geographic information systems (GIS) allows for spatial analysis of water quality data, enhancing the understanding of how local conditions and human activities influence water quality across different regions. This spatial awareness can guide resource allocation and prioritize areas that require urgent attention or intervention [29, 30].

Additionally, the ability to analyze data in real time supports community engagement and transparency. Stakeholders, including local governments and environmental organizations, can share water quality information with the public, fostering awareness and encouraging community involvement in conservation efforts [31–34]. This collaborative approach not only enhances public trust but also empowers citizens to take part in monitoring and protecting their water resources.

9.3.3 SUPPORT FOR ENVIRONMENTAL PROTECTION

By predicting potential pollution events or water quality degradation, AI can play a crucial role in environmental conservation efforts. This predictive capability allows for early detection of changes in water quality that may indicate the onset of harmful events, such as nutrient runoff leading to algal blooms or toxic contaminant spills. By identifying these threats before they escalate, AI-driven systems enable stakeholders to implement timely and effective responses that mitigate environmental damage.

One of the key benefits of such proactive management measures is the protection of aquatic ecosystems, which are often sensitive to changes in water quality [35]. For example, when predictive algorithms forecast a risk of hypoxia due to falling dissolved oxygen levels, conservationists can take immediate actions, such as managing nutrient inputs from agriculture or initiating aeration techniques in affected water bodies. This not only helps sustain fish populations and other marine life but also preserves the overall health of the ecosystem, which is vital for maintaining biodiversity [36].

Moreover, AI's ability to analyze long-term data trends helps identify and target specific areas at higher risk for future pollution events. By using historical data in tandem with real-time monitoring, AI models can pinpoint patterns in water quality degradation, thereby informing land-use planning and environmental policy decisions. For instance, if certain watersheds consistently show elevated pollutant levels during specific seasons, policymakers can develop strategies tailored to these periods, such as implementing stricter regulations on runoff during high-risk times.

Additionally, AI can enhance collaboration across various stakeholders, including governmental agencies, non-profit organizations, and local communities. By providing a comprehensive and data-driven understanding of water quality dynamics, AI facilitates integrated resource management approaches that involve multiple parties. This collaborative focus allows for streamlining efforts in conservation initiatives, making them more effective and efficient.

9.3.4 PUBLIC HEALTH SAFEGUARDS

AI-driven predictions can enhance the safety of drinking water supplies by identifying contamination risks early. This proactive approach allows utilities and health officials to monitor water quality continuously and detect potential threats before they reach harmful levels. By leveraging machine learning algorithms that analyze historical data, real-time sensor inputs, and environmental factors, AI systems can recognize patterns indicative of contamination events, such as sudden changes in chemical composition or the presence of pathogens [37–39].

Early identification of contamination risks leads to timely alerts and interventions that are crucial for protecting public health and preventing waterborne illnesses. For example, if an AI system detects unusual spikes in bacteria levels or chemical pollutants, it can immediately notify water treatment facilities to implement corrective measures, such as increased filtration or disinfection processes. This rapid response capability is vital, especially in urban areas where large populations rely on a single water source [40].

Moreover, AI-driven predictions can help prioritize testing and monitoring efforts in high-risk areas. By analyzing factors such as geographic location, historical contamination data, and weather patterns, AI can identify regions that are more susceptible to water quality issues. This targeted approach allows water utilities to allocate resources more effectively, ensuring that vulnerable populations receive the protection they need.

In addition to immediate alerts, AI systems can also provide long-term insights into trends and potential future risks. By forecasting seasonal variations in water quality or predicting the impact of urban development and agricultural practices on local water sources, these systems can inform strategic planning and decision-making. For instance, if predictions indicate a higher likelihood of contamination during specific weather events, water managers can take preemptive actions, such as adjusting water treatment protocols or increasing public awareness campaigns during those times [41].

Furthermore, the integration of AI with public health data can enhance the understanding of the relationship between water quality and health outcomes [42]. By correlating water quality predictions with epidemiological data, health officials can identify potential outbreaks of waterborne diseases more swiftly and accurately. This collaboration between environmental monitoring and public health surveillance ensures a comprehensive approach to safeguarding community health.

9.4 CHALLENGES AND CONSIDERATIONS

While AI holds tremendous promise for water quality control, challenges persist. As in every domain where artificial intelligence is used, data and data quality hold a key value to the success or the accuracy of the models, the study [43] highlights the critical importance of data quality and data privacy in artificial intelligence (AI) applications, noting challenges such as inaccurate, biased, or unreliable data affecting AI outcomes and privacy breaches leading to legal and ethical issues. Strategies like data cleansing, bias mitigation, anonymization, encryption, and collaboration with stakeholders are proposed to address these challenges. Establishing robust data governance frameworks, ensuring transparency, and adhering to regulatory compliance are emphasized as essential steps in building ethical and responsible AI systems capable of maintaining user trust and privacy. Biswaranjan Acharya [44] addresses the challenges of implementing a standard regulatory framework for artificial intelligence (AI). It discusses various legal mechanisms to mitigate public risks associated with AI. Despite difficulties in providing an exact definition for AI, the paper suggests that precise legal definitions are possible. Torts and contracts are explored as legal mechanisms, with an emphasis on liability in cases of civil wrongs or voluntarily accepted duties. Product defects are covered under tort law and consumer protection laws, allowing for claims against AI system owners or users. As the field continues to evolve, emerging trends and future directions have come to the forefront. Integrating AI with emerging technologies, such as the Internet of Things (IoT) and edge computing, holds the potential for creating robust and decentralized water quality monitoring systems. Addressing environmental justice concerns and incorporating community-driven approaches in AI-based water quality initiatives are also gaining prominence. Artificial intelligence ethics also come into consideration. Pant et al. [45] focused their study on AI practitioners' awareness of AI ethics and challenges in its implementation, based on a survey of 100 practitioners. It reveals a moderate awareness of AI ethics, primarily influenced by workplace rules, with privacy and security being the most recognized principles. Challenges include biases, technology complexities, and aligning AI with human values, urging for further research and guidelines for ethical AI development.

Socially, Liu [46] emphasizes the need to consider AI's societal impact in life sciences for sustainable and stable research. They also discussed concerns about AI's potential to replace human jobs in sectors like production and driving but also highlighted optimism about creating new job opportunities. AI is seen as a force for driving societal productivity and creating wealth, but also arise questions about ensuring fair distribution of benefits to prevent job displacement and inequality. Balancing AI's economic benefits with social equity is crucial for its long-term impact on society.

9.5 CONCLUSION

In conclusion, the integration of artificial intelligence (AI) in water quality prediction represents a pivotal advancement toward sustainable water resource management and environmental conservation. The utilization of diverse AI techniques, including machine learning algorithms and neural network models, has demonstrated remarkable efficacy in accurately forecasting water quality parameters and indices. These predictive models offer valuable insights for decision-makers, empowering them to anticipate and address potential pollution events, mitigate risks, and enact effective interventions. However, challenges such as data quality, privacy concerns, regulatory frameworks, and ethical considerations persist. Addressing these challenges is paramount to ensuring the responsible and equitable deployment of AI in water quality control initiatives. Looking ahead, integrating AI with emerging technologies and adopting community-driven approaches will further enhance the effectiveness and inclusivity of AI-based water quality initiatives, fostering sustainable development and societal well-being.

REFERENCES

[1] M. Koranga, P. Pant, T. Kumar, D. Pant, A. K. Bhatt, and R. P. Pant, "Efficient Water Quality Prediction Models Based on Machine Learning Algorithms for Nainital Lake, Uttarakhand," *Materials Today: Proceedings*, vol. 57, pp. 1706–1712, Jan. 2022. https://doi.org/10.1016/j.matpr.2021.12.334.

[2] U. Ahmed, R. Mumtaz, H. Anwar, A. A. Shah, R. Irfan, and J. García-Nieto, "Efficient Water Quality Prediction Using Supervised Machine Learning," *Water*, vol. 11, no. 11, Art. no. 11, Nov. 2019. https://doi.org/10.3390/w11112210.

[3] R. Rana et al., "Artificial Intelligence for Surface Water Quality Evaluation, Monitoring and Assessment," *Water*, vol. 15, no. 22, Art. no. 22, Jan. 2023. https://doi.org/10.3390/w15223919.

[4] L. Chen, T. Wu, Z. Wang, X. Lin, and Y. Cai, "A Novel Hybrid BPNN Model Based on Adaptive Evolutionary Artificial Bee Colony Algorithm for Water Quality Index Prediction," *Ecological Indicators*, vol. 146, p. 109882, Feb. 2023. https://doi.org/10.1016/j.ecolind.2023.109882.

[5] X. Yan, T. Zhang, W. Du, Q. Meng, X. Xu, and X. Zhao, "A Comprehensive Review of Machine Learning for Water Quality Prediction over the Past Five Years," *Journal of Marine Science and Engineering*, vol. 12, no. 1, Art. no. 1, Jan. 2024. https://doi.org/10.3390/jmse12010159.

[6] M. Kulisz and J. Kujawska, "Application of Artificial Neural Network (ANN) for Water Quality Index (WQI) Prediction for the River Warta, Poland," *Journal of Physics: Conference Series*, vol. 2130, no. 1, p. 012028, Dec. 2021. https://doi.org/10.1088/1742-6596/2130/1/012028.

[7] Z. Z. Aljanabi, A.-H. M. J. Al-Obaidy, and F. M. Hassan, "A Novel Water Quality Index for Iraqi Surface Water," *Baghdad Science Journal*, vol. 20, no. 6(Suppl.), p. 2395, Dec. 2023. https://doi.org/10.21123/bsj.2023.9348.

[8] N. Mamat, S. F. Mohd Razali, and F. B. Hamzah, "Enhancement of Water Quality Index Prediction Using Support Vector Machine with Sensitivity Analysis," *Frontiers in Environmental Science*, vol. 10, Jan. 2023. https://doi.org/10.3389/fenvs.2022.1061835.

[9] J. Mohd Zebaral Hoque, N. A. Ab. Aziz, S. Alelyani, M. Mohana, and M. Hosain, "Improving Water Quality Index Prediction Using Regression Learning Models," *International Journal of Environmental Research and Public Health*, vol. 19, no. 20, Art. no. 20, Jan. 2022. https://doi.org/10.3390/ijerph192013702.

[10] S. L. Chia, M. Y. Chia, C. H. Koo, and Y. F. Huang, "Integration of Advanced Optimization Algorithms Into Least-Square Support Vector Machine (LSSVM) for Water Quality Index Prediction," *Water Supply*, vol. 22, no. 2, pp. 1951–1963, Sep. 2021. https://doi.org/10.2166/ws.2021.303.

[11] J. P. Nair and M. S. Vijaya, "River Water Quality Prediction and Index Classification Using Machine Learning," *Journal of Physics: Conference Series*, vol. 2325, no. 1, p. 012011, Aug. 2022. https://doi.org/10.1088/1742-6596/2325/1/012011.

[12] F. C. Mbachu and I. L. Nwaogazie, "Modelling and Prediction of Water Quality Index of Selected Pond Water in Aboh Mbaise Local Government Area, Imo State, Nigeria," *JERR*, vol. 25, no. 11, pp. 13–21, Nov. 2023. https://doi.org/10.9734/jerr/2023/v25i111016.

[13] S. Vijay and K. Kamaraj, "Prediction of Water Quality Index in Drinking Water Distribution System Using Activation Functions Based Ann," *Water Resour Manage*, vol. 35, no. 2, pp. 535–553, Jan. 2021. https://doi.org/10.1007/s11269-020-02729-8.

[14] B. Singh, P. Sihag, V. P. Singh, A. Sepahvand, and K. Singh, "Soft Computing Technique-Based Prediction of Water Quality Index," *Water Supply*, vol. 21, no. 8, pp. 4015–4029, Jun. 2021. https://doi.org/10.2166/ws.2021.157.

[15] A. Derdour et al., "Prediction of Groundwater Quality Index Using Classification Techniques in Arid Environments," *Sustainability*, vol. 15, no. 12, Art. no. 12, Jan. 2023. https://doi.org/10.3390/su15129687.

[16] G. Volf, I. Sušanj Čule, E. Žic, and S. Zorko, "Water Quality Index Prediction for Improvement of Treatment Processes on Drinking Water Treatment Plant," *Sustainability*, vol. 14, no. 18, Art. no. 18, Jan. 2022. https://doi.org/10.3390/su141811481.

[17] A. E. Alprol, A. T. Mansour, M. E. E.-D. Ibrahim, and M. Ashour, "Artificial Intelligence Technologies Revolutionizing Wastewater Treatment: Current Trends and Future Prospective," *Water*, vol. 16, no. 2, p. 314, 2024.

[18] S. Talukdar et al., "Optimisation and Interpretation of Machine and Deep Learning Models for Improved Water Quality Management in Lake Loktak," *Journal of Environmental Management*, vol. 351, p. 119866, Feb. 2024. https://doi.org/10.1016/j.jenvman.2023.119866.

[19] N. Lansbury, "Preventing Disease Through Healthy Environments: The Contribution of Environmental Health," in *Handbook of Concepts in Health, Health Behavior and Environmental Health*, Springer, 2024, pp. 1–20.

[20] J. Mabrouki, M. Azrour, and S. El Hajjaji, "Use of Internet of Things for Monitoring and Evaluation Water's Quality: Comparative Study," *International Journal of Cloud Computing*, vol. 10, no. 5–6, pp. 633–644, 2021.

[21] J. Mabrouki et al., "Study, Simulation and Modulation of Solar Thermal Domestic Hot Water Production Systems," *Modeling Earth Systems and Environment.*, Jun. 2021. https://doi.org/10.1007/s40808-021-01200-w.

[22] M. Azrour, J. Mabrouki, G. Fattah, A. Guezzaz, and F. Aziz, "Machine Learning Algorithms for Efficient Water Quality Prediction," *Modeling Earth Systems and Environment*, vol. 8, no. 2, pp. 2793–2801, 2022.

[23] F. Mousli, J. Mabrouki, L. Bouhachlaf, M. Azrour, and S. E. Hajjaji, "Detection of Some Water Elements Based on IoT: Review Study," *IoT and Smart Devices for Sustainable Environment*, Springer Nature, 2022, pp. 1–17.

[24] A. Anouzla et al., "Multi-Response Optimization of Coagulation–Flocculation Process for Stabilized Landfill Leachate Treatment Using a Coagulant Based on an Industrial Effluent," *Desalination and Water Treatment*, p. 10, 2022.

[25] J. Mabrouki, M. Azrour, and S. E. Hajjaji, "Use of Internet of Things for Monitoring and Evaluating Water's Quality: A Comparative Study," *International Journal of Cloud Computing*, vol. 10, no. 5–6, pp. 633–644, 2021.

[26] J. Mabrouki, G. Fattah, S. Kherraf, Y. Abrouki, M. Azrour, and S. El Hajjaji, "Artificial Intelligence System for Intelligent Monitoring and Management of Water Treatment Plants," in *Emerging Real-World Applications of Internet of Things*, CRC Press, 2022, pp. 69–87.

[27] J. Mabrouki et al., "Geographic Information System for the Study of Water Resources in Chaâba El Hamra, Mohammedia (Morocco)," in *Artificial Intelligence and Smart Environment: ICAISE'2022*, Springer, 2023, pp. 469–474.

[28] J. Mabrouki, M. Azrour, A. Boubekraoui, and S. El Hajjaji, "Simulation and Optimization of Solar Domestic Hot Water Systems," *International Journal of Social Ecology and Sustainable Development*, vol. 13, no. 1, 2022. https://doi.org/10.4018/IJSESD.315309.

[29] B. Kuhaneswaran, G. Chamanee, and B. T. G. S. Kumara, "A Comprehensive Review on the Integration of Geographic Information Systems and Artificial Intelligence for Landfill Site Selection: A Systematic Mapping Perspective," *Waste Management & Research*, p. 0734242X241237100, Apr. 2024. https://doi.org/10.1177/0734242X241237100.

[30] P. Jayaraman, K. K. Nagarajan, P. Partheeban, and V. Krishnamurthy, "Critical Review on Water Quality Analysis Using IoT and Machine Learning Models," *International Journal of Information Management Data Insights*, vol. 4, no. 1, p. 100210, Apr. 2024. https://doi.org/10.1016/j.jjimei.2023.100210.

[31] M. Mohy-Eddine, M. Azrour, J. Mabrouki, F. Amounas, A. Guezzaz, and S. Benkirane, "Embedded Web Server Implementation for Real-Time Water Monitoring," in *Advanced Technology for Smart Environment and Energy*, J. Mabrouki, A. Mourade, A. Irshad, and S. A. Chaudhry, Eds., in Environmental Science and Engineering, Cham: Springer International Publishing, 2023, pp. 301–311. https://doi.org/10.1007/978-3-031-25662-2_24.

[32] M. Azrour, J. Mabrouki, A. Guezzaz, S. Benkirane, and H. Asri, "Implementation of Real-Time Water Quality Monitoring Based on Java and Internet of Things," in *Integrating Blockchain and Artificial Intelligence for Industry 4.0 Innovations*, S. Goundar and R. Anandan, Eds., in EAI/Springer Innovations in Communication and Computing, Cham: Springer International Publishing, 2024, pp. 133–143. https://doi.org/10.1007/978-3-031-35751-0_8.

[33] S. Nasrdine, M. Benchrifa, N. Ben-Lhachemi, J. Mabrouki, M. Slaoui, and M. Azrour, "New Design of an Inclined Solar Distiller for Freshwater Production: Experimental Study," in *Technical and Technological Solutions Towards a Sustainable Society and Circular Economy*, J. Mabrouki and A. Mourade, Eds., Cham: Springer Nature Switzerland, 2024, pp. 447–453. https://doi.org/10.1007/978-3-031-56292-1_36.

[34] M. Azrour et al., "A Survey of Machine and Deep Learning Applications in the Assessment of Water Quality," *World Sustainability Series*, vol. Part F2854, 2024, pp. 471–483. https://doi.org/10.1007/978-3-031-56292-1_38.

[35] A. Viani, T. Orusa, E. Borgogno-Mondino, and R. Orusa, "A One Health Google Earth Engine Web-GIS Application to Evaluate and Monitor Water Quality Worldwide," *Euro-Mediterranean Journal for Environmental Integration*, May 2024. https://doi.org/10.1007/s41207-024-00528-w.

[36] L. Wang, "Advances in Monitoring and Managing Aquatic Ecosystem Health: Integrating Technology and Policy," *International Journal of Aquaculture*, vol. 14, no. 0, Art. no. 0, Apr. 2024, Accessed: Nov. 18, 2024. [Online]. Available: https://aqua-publisher.com/index.php/ija/article/view/3861.

[37] C. H. Pérez-Beltrán, A. D. Robles, N. A. Rodriguez, F. Ortega-Gavilán, and A. M. Jiménez-Carvelo, "Artificial Intelligence and Water Quality: From Drinking Water to Wastewater," *TrAC Trends in Analytical Chemistry*, vol. 172, p. 117597, Mar. 2024. https://doi.org/10.1016/j.trac.2024.117597.

[38] S. Dargaoui et al., "Applications of Blockchain in Healthcare: Review Study," in *IoT, Machine Learning and Data Analytics for Smart Healthcare*, CRC Press, 2024.

[39] S. Dargaoui et al., "IoT-Driven Smart Agriculture: Security Issues and Authentication Schemes Classification," in *International Conference on Connected Objects and Artificial Intelligence*, Cham: Springer Nature Switzerland, 2024, pp. 61–66.

[40] C. Chellaiah et al., "Integrating Deep Learning Techniques for Effective River Water Quality Monitoring and Management," *Journal of Environmental Management*, vol. 370, p. 122477, Nov. 2024. https://doi.org/10.1016/j.jenvman.2024.122477.

[41] A. Mandal and A. R. Ghosh, "Role of Artificial Intelligence (AI) in Fish Growth and Health Status Monitoring: A Review on Sustainable Aquaculture," *Aquaculture International*, vol. 32, no. 3, pp. 2791–2820, Jun. 2024. https://doi.org/10.1007/s10499-023-01297-z.

[42] K. Javan, A. Altaee, S. BaniHashemi, M. Darestani, J. Zhou, and G. Pignatta, "A Review of Interconnected Challenges in the Water–Energy–Food Nexus: Urban Pollution Perspective towards Sustainable Development," *Science of The Total Environment*, vol. 912, p. 169319, Feb. 2024. https://doi.org/10.1016/j.scitotenv.2023.169319.

[43] K. Yekaterina, "Challenges and Opportunities for AI in Healthcare," *International Journal of Law and Policy*, vol. 2, no. 7, Art. no. 7, Jul. 2024. https://doi.org/10.59022/ijlp.203.

[44] B. Acharya, K. Garikapati, A. Yarlagadda, and S. Dash, "Chapter 1 – Internet of Things (IoT) and Data Analytics in Smart Agriculture: Benefits and Challenges," in *AI, Edge and IoT-Based Smart Agriculture*, A. Abraham, S. Dash, J. J. P. C. Rodrigues, B. Acharya, and S. K. Pani, Eds., in Intelligent Data-Centric Systems, Academic Press, 2022, pp. 3–16. https://doi.org/10.1016/B978-0-12-823694-9.00013-X.

[45] A. Pant, R. Hoda, S. V. Spiegler, C. Tantithamthavorn, and B. Turhan, "Ethics in the Age of AI: An Analysis of AI Practitioners' Awareness and Challenges," *ACM Transactions on Software Engineering and Methodology*, vol. 33, no. 3, pp. 80:1–80:35, Mar. 2024. https://doi.org/10.1145/3635715.

[46] T. Liu and W. Li, "Applications and Challenges of Artificial Intelligence in Life Sciences," *SHS Web of Conferences*, vol. 187, p. 04007, 2024. https://doi.org/10.1051/shsconf/202418704007.

10 Aspects of IoT-Based Applications Intended for Transforming the World on a Large Scale

Shafqat Ul Ahsaan, Ashish Kumar Mourya, and Seema Rani

10.1 INTRODUCTION

Artificial intelligence (AI) is not a new term. However, AI-based applications are making way in every imaginable sector of the global economy. AI can simply be defined as a system that imitates human intelligence and behavior. Like humans, it learns from the data which is fed into it from multiple sources, using IoT or any of its supporting associative technology. It uses neural networks; machine learning (deep learning techniques) to develop results from the data it processes. Today, AI-based applications are widely used in every smart device and innovative technology like big data, cloud computing, IoT, and so on. Needless to say, the rate at which AI is progressing, means we need to wisely monitor its development process, so that it does not become an unwanted liability [1].

IoT devices have the capacity to interact with each other. Nowadays, we are almost surrounded by IoT devices which are communicating through multiple devices. IoT has also gained interest in academics, researchers, and entrepreneurs. IoT goes through sensing, processing, and communicating processes without any human intervention. IoT has also become popular through an idea of developing smart homes, smart cities, and other intelligent things. It also provides opportunities and development as it has been raised to approximately 50 billion connected devices by the end of 2020. IoT devices are also gaining popularity in industries and business. IoT has played an important role in conserving the energy from small scale to the large scale. The concept of IoT has also introduced automation to reduce manpower and increase the efficiency of the work [2].

IoT devices have sensors that can sense the environment by collecting the required information from the environment and then monitoring output through different dashboards. At last it requires a device with the capability of serving and routing. Some of the sensors that are commonly used in these IoT devices are accelerometers, temperature sensors, proximity sensors, gyroscopes, image sensors, and so on. But one of the best examples of an IoT device is the smartphone. Smartphones can be

DOI: 10.1201/9781003527664-10

connected through smart watches and smart bands to trace activity, to collect data, and maintain records of an individual [3, 4].

IoT has various roles to play in different domains:

- *Healthcare Applications:* There are several devices that can be operated through the technology being used in IoT. Smart watches and fitness bands provide an opportunity to monitor and track health at regular intervals.
- *Energy Applications:* There is a rise in the consumption of energy in every sector, be it a household or for an industrial purpose. So, to monitor and track the consumption, IoT is becoming a need for today's world.
- *Education:* IoT has identified and worked on the gaps of present education systems. It enhances the quality of education while reducing the cost.
- *Air and Water Pollution:* IoT uses various sensors to detect and monitor the pollution in the air and water. It helps by providing proper tracking of the contamination of air and water.
- *Farming:* It reduces human intervention in continuous monitoring of farmlands, including the quality of crops to be produced, soil, environment, and so on.
- *Transportation:* IoT has changed the transportation sector. We are now being introduced to the self-driving cars that require sensors to track the path provided, monitor the traffic, sense the traffic signals, parking assistance to provide the location of free parking slots.
- *Government:* Governments are trying to build smart cities to improve the work–life balance of an individual. IoT is also being introduced in the army to provide better security across borders.

10.2 CHALLENGES AND ISSUES FACED BY IoT

IoT has gained popularity in almost every field now. Increasing the number of IoT devices has impacted and increased the number of people joining the chain of IoT. This can ultimately lead to the vulnerability of the security of these devices, data present and transmitted through these devices. So, IoT security has become a major concern of IT world. It is crucial to ensure security of the networks with connected IoT devices. IoT is a broad area and IoT security is even broader [5, 6].

IoT devices have a larger attack surface due to their Internet-supported connectivity. This provides opportunities for hackers to interact with devices. This is why phishing is becoming more and more common these days. To overcome these issues, cloud security has been introduced. Attacks on IoT devices are also increasing due to the negligence of industries. Many firms are introducing IoT in most of their work today; IoT devices are commonly used in healthcare and automotive companies, but these companies are not investing the amount of money necessary to secure these devices [7]. This will ultimately lead to exposing the devices and their data to the outside world increasing cybersecurity attacks.

Another major issue faced by IoT security is limited resources of these devices. Many of these devices are not designed for the complex structure of firewalls and antivirus software [8], which can then ultimately lead to data breaches. These can be protected by introducing IoT security during the designing phase, adding PKI security and API security. Many privacy issues by customers that devices collect their activity on smart devices that can lead to potential attacks. So, this is why blockchain needs to be introduced [9, 10].

As the demand for IoT has been emerging, the problems concerning various factors like security and privacy have been increasing. Blockchain has brought many benefits in an era of IoT. Blockchain technology can provide better solutions to the problems faced by the IoT systems [11].

10.3   TRANSFORMATION OF THE HEALTH SECTOR WITH IoT

In today's healthcare landscape, the necessity for change looms large, spurred by a convergence of factors such as technological advancements, evolving patient requirements, and economic pressures. Transformational efforts extend beyond mere adoption of new technologies; they involve a fundamental reshaping of healthcare delivery to improve accessibility, affordability, and quality of care. A key catalyst for this transformation is the integration of computerized technologies. Telemedicine technologies such as wearable devices, EHRs, and artificial intelligence (AI) are reshaping healthcare by facilitating remote consultations, simplifying administrative tasks, and tailoring treatment plans. These advancements empower patients to play a more active role in their health management while enabling providers to offer more efficient care [12, 13].

Furthermore, healthcare transformation involves a shift toward value-based care models that prioritize outcomes and patient satisfaction over service volume. This approach fosters collaboration among providers, emphasizes preventive care and chronic disease management, and utilizes data analytics to enhance resource allocation. Patient-centered care is another crucial aspect, involving customized treatment plans, open communication, and holistic health approaches that address social determinants. Moreover, fostering innovation and collaboration across the healthcare ecosystem is essential. This includes partnerships between healthcare organizations, technology firms, academic institutions, and government agencies to develop and implement cutting-edge solutions. In essence, healthcare transformation demands vision, leadership, and collaboration. By embracing digital technologies, transitioning to value-based care, prioritizing patient-centered approaches, and fostering innovation, organizations can drive meaningful change toward a more accessible, affordable, and equitable healthcare system [14].

10.4   AGRICULTURAL IoT SYSTEM ARCHITECTURE

The Internet of Things provides various solutions across different sectors, including healthcare, retail, traffic management, security, connected homes, urban infrastructure, and farming. The deployment of IoT in agriculture is seen as the perfect solution due to the continuous need for monitoring and controlling in this field. The primary uses of IoT in agriculture are site-specific crop management, domestic animals, and greenhouses, which constitute various monitoring categories. It is a network where physical objects such as florae, faunae, production tools, and elements related to environment, and other essential entities that are part of agriculture system are linked with the Internet via farming information awareness kit using definite procedures for data exchange [14, 15]. All these applications are effectively monitored through the utilization of diverse IoT-based sensors and devices using wireless sensor networks (WSNs), enabling farmers to seamlessly gather pertinent data via sensing devices. This technology can be utilized for identifying the best time for harvesting, matching farmers with suitable conditions, identifying diseases, and controlling machinery. Certain Internet of Things (IoT) configurations use cloud services to analyze and handle data from remote locations. This technology assists researchers and farmers in making more informed choices by making agricultural objects and processes detectible, positioned, followed, and monitored rationally [16, 17]. It can also aid with the ability to govern complicated farming systems, help in management for agricultural disasters, and considerably improve the ability of humans to understand the distinct features of crops and animals.

The use of IoT technology across the globe, especially in agriculture sciences, has been made possible by the advancement of the Internet and computational technology. As a result, sensors manufactured with advanced technologies are continuously evolving and moving in the direction of becoming small, smart, and unified [18]. In terms of sensor technology and industrial methods, the United States, Japan, and Germany currently hold a leading position over other nations. The soil, weather, water, and plant sensors are among many diverse uses for farming, and are becoming more and more common. The sensors that identify different items offer significant assistance in gathering data on agricultural productivity [19].

The agricultural Internet of Things system's design and implementation is mostly based on its system architecture. Researchers have undertaken in-depth studies on agricultural IoT and have put forth a number of architectures.

Two IoT architectural initiatives were established under the EU's Seventh Framework program (Figure 10.1). One of the projects is SENSEI, which considers the Internet as an infrastructure that links the physical and digital realms [20]. In this project, Radio Frequency Identification (RFID), wireless sensing, and execution networks all are intended to be integrated in order to create a real-world Internet structure that is open to business and may offer services through a single interface. The next project, called IoT A, aims to describe the essential elements of IoT and create a reference structural model for it [21, 22]. Rather than representing the structure of a particular application, the reference frame working model is a generalization of the Internet of Things workings. As such, it offers the finest illustration on various application domains to create more interoperable IoT architectures. The IoT architecture has been divided into three layers. The two primary shortcomings of this division approach are that it is unable to reflect the features of individual users as well as depict the qualities and modifications of IoT technology in certain engineering practices. In order to address this issue, we separate the farming IoT system design into five layers: the user layer, application layer, transport layer, perception layer, and object layer, as depicted in Figure 10.1. In order to examine agricultural IoT architecture, one should conduct the subsequent

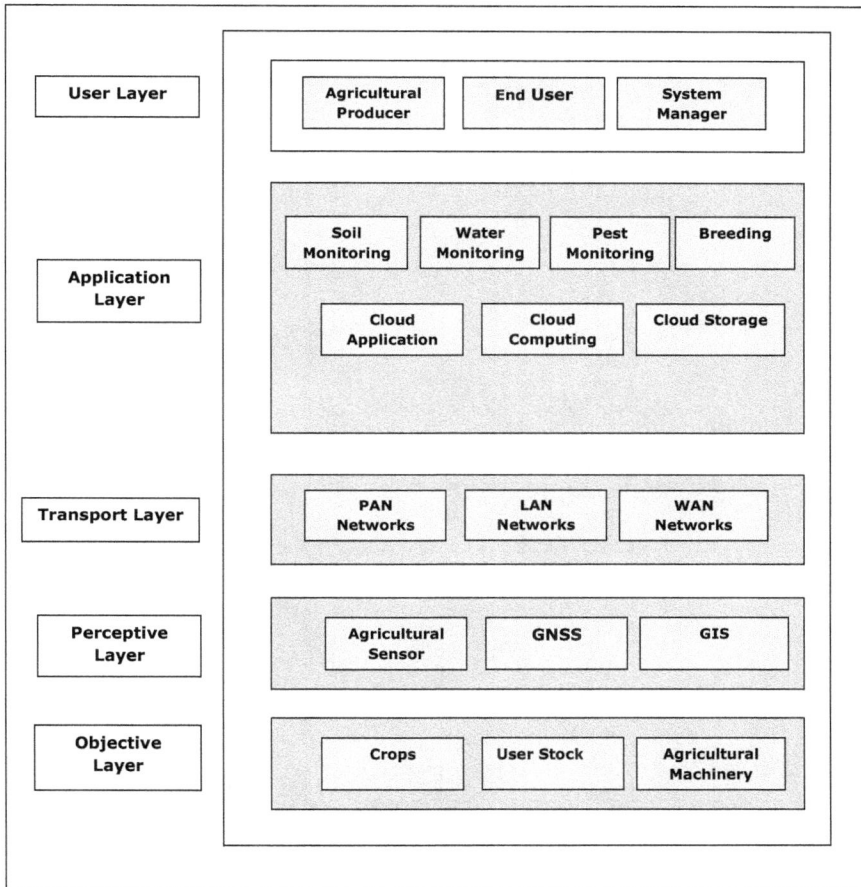

FIGURE 10.1   Framework for smart farming.

actions: (i) IoT applications and situations should be abstracted. (ii) Outline the overall requirements and guiding principles of the Internet of Things architecture. (iii) Define the overall functional structure of the Internet of Things and further subdivide its core architecture [23].

10.5   IoT APPLICATIONS IN CROP MANAGEMENT

The Zigbee wireless network of agricultural IoT permits wireless self-organized data transmission. It certifies stable and convenient remote data transmission through efficient integration with wired data transmission. European and American nations have been using satellites to perform accurate field cultivation operations and monitoring, as well as intelligent fertilizer and water monitoring, in real time for agricultural output [24, 25]. The integration of artificial technologies can enhance the usage of

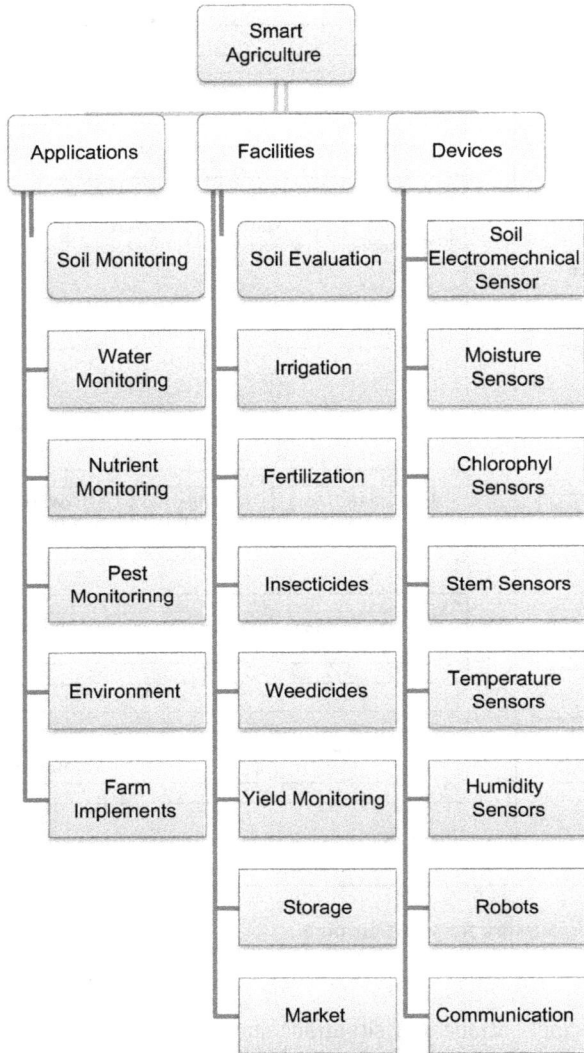

FIGURE 10.2 Application of IoT in smart farming.

sensor data through intelligent administration and monitoring (Figure 10.2). When coupled with expert systems, agricultural IoT enables precise crop management and improves the planting experience for farmers [26].

IoT has been employed in China in a wide range of agricultural fields, including grazing, planting, and animal farming. In agricultural production, it has also been utilized for environmental monitoring, ensuring the safety traceability of agricultural products and irrigating farmland. Furthermore, China has created highly accurate information monitoring and analysis tools, which has encouraged the use of IoT in agriculture. Currently, equipment for gathering data on crops and plants, tracking environmental data, and observing animal behavior make up the majority of the equipment that have been developed and used.

10.6 BLOCKCHAIN WITH IoT (BIoT)

The Internet of Things connects the devices, objects, and goods to provide opportunities to capture data from sensors and actuators in the environment. These data are then transmitted to a centralized server, often a cloud server. These IoT devices are now widely used in large firms and business. So, the privacy concerns of these devices are growing speedily. The attacks that generally occur on IoT devices are DDoS attacks, Ransomware attacks, and malicious attacks. DDoS attacks are those where the central server is bombarded with many data requests confusing the server and resulting in denial of service for authenticated requests. Furthermore, when the IoT ecosystem becomes more broad, the problem of authenticating, approving, and connecting new devices increases with the present centralized server.

So, to solve this IoT problem, a blockchain technology, known as Distributed Transaction Ledge (DTL), has to be addressed. DTL technology is a tamper-proof mechanism which removes the need of the central party to recognize the authenticated devices. The deployment of blockchain in IoT has brought a new domain called BIoT. With BIoT, no single server has control over the data generated by IoT devices. Blockchain with IoT gives devices the opportunity to track past transactions which leads to clearing tracks of blocks. Thus, this will help trace any data leaked and can be redeemed [27, 28].

10.6.1 APPLICATIONS OF INTERNET OF THINGS WITH BLOCKCHAIN IN VARIOUS DOMAINS

Similar to IoT, blockchain also has a wider area of applications. Dividing on the basis of the applications, we can have a sector-based application and a category-based application. Let's look into the sector-based application of the blockchain in Table 10.1.

Blockchain has various other applications based on the category they have been in use. Now let's look into the category-based application of blockchain technology in Table 10.2.

TABLE 10.1
Blockchain in Different Sectors

Sector	Applications
Healthcare	Electronic medical records, digital prescription records, digital case reports, digital previous medical records
Smart city	Smart data maintenance, smart transactions, smart pollution control, digital data, energy management, water management
Agriculture	Processing records related to agriculture, soil data, digitalized growth rate
Business	Import and export of data, digital records of data, digital financing, data transactions
Energy	Resource availability, demand data records, tracking of resources, energy raw material data
Finance	Currency exchanges, money deposits, money transfers, smart contracts, cryptocurrency
Transport	Transport records, toll maintenance, vehicle tracking, good delivery
Others	Digital content, economy sharing, ownership, government and voting, virtual nations

TABLE 10.2
Blockchain Application in Different Categories

Category	Applications
Financial transactions	It includes records of various activities like mutual funds, stocks, Insurance, bonds, and pensions
Currency	It is used in the process of creating a digital currency like bitcoin and cryptocurrency
Identification	It is used in recognizing any identity like verification of a passport, identity card, voting card, and so on
Physical assets key	It is used to secure the information of the key that accesses the physical assets of the individual that needs to be protected, for example, offices, lockers, and so on
Intangible assets	These assets do not have any physical existence in nature, for example, trademarks and brand recognition
General	This includes third-party attribution, multiparty signatures, and transactions

10.7 BIG DATA ANALYTICS WITH IoT

Recording of data through various means and methods has been in practice since the dawn of our civilization. But in the present era of IoT, in 2019, as per the results published by Statista Research Department, it is estimated that there are 26.66 billion IoT-based connected devices around the globe, and it is expected to cross 75.44 billion marks by 2025. As per another cloud-based operating system company Domo, by 2020 there will be 40 times more data than there are in the observable universe. As per one previous report of Ralph Jacobson, more than 90% of the data of human civilization has come into existence in the last couple of years [29].

Our simple smart watches, mobile phones and mobile apps, cars, televisions, laptops, desktops, railways, aviation, telecommunication, hospital records, banks, closed circuit cameras, e-commerce sites, social sites, whether sensors, satellites, and so on, all are recording some form of data in various varieties, be it simple text, audio, videos, and so on. Besides, with the availability of broadband and IoT-enabled platforms, such voluminous data is being stored and shared at a never-before velocity and with some organized protocols it has the quality of verifiable veracity.

Now where does all this data go or where is it stored? Most common technique in IoT platforms is to store it on cloud for the purpose of its further analysis. We know that the dots can only be connected to the past; nevertheless, if appropriate analytics is applied, some meaningful patterns and knowledge can be deduced and developed from this gigantic amount of data, which is getting bigger and bigger with every passing second.

In the modest terms, the phrase big data is used to denote such datasets, which are beyond the capacity of traditional databases in terms of storage, processing, and manipulation. Big data analytics is presumed to possess some basic features like heavy volume of data, lightning velocity computational capacities, and a wide degree of data on its variety aspects. This gigantic amount of data, which are stored on cloud servers and manipulated through cloud computing operating systems in an IoT-based infrastructure, acts as the basic fuel for big data analytics techniques, which are advanced enough to handle data that may be manipulated from its unstructured form to semistructured and finally to a structured form. So the meaningful configurations and patterns can be derived from them and value-added information may be generated for optimization of the system as well as improving the quality of service to the end users.

In a big data analytics process, when raw data is input, configuring its volume, variety, veracity, and value is treated with two types of algorithms, namely, dynamic algorithms (using machine learning and AI) and static algorithms (expert systems). It results into value-added information and KNOWLEDGE datasets, which can further be applicable in improved human decision-making or automated decision-making and may even be used by personal assistant system to provide better quality of services.

In a big data ecosystem, IoT and blockchain continuously carry on feeding it with more real-time generated secured data streams through peer-to-peer, peer-to-machine, and machine-to-machine networking. Besides, the variety of fields in which big data is being used, it has promising potentials like enhanced transparency, predictive analysis, optimized processes, increased efficiency, improved customer profiling and preferences, and customer-specific tailor-made products and strategies.

Nowadays, more than ever businesses and corporate houses are in inevitable need of user's data, to analyze, evaluate, and predict new strategies for not only surviving in the global market but also to serve their users better and thereby keeping an edge above one another. Briefly, every big data process primarily undergoes stages of acquiring of data, cleaning the data on the basis of its specifications, and preparing it for analysis through advanced pattern-seeking algorithms and procedures for deriving information and knowledge from it. And finally, to visualize the data in form of graphs, charts, and tables for decision-making in reference to the patterns, needs, and predictions for the end user. Summarily, the analytical challenges of big data are also of basically three categories related to data's volume, variety, value, veracity, and velocity; processing of data in reference to its acquisition, cleaning, and integration; and finally, management of data's security, privacy, data governance, sharing of data and information, operational cost, and ownership issues of data [30].

10.8   CLOUD COMPUTING AND IoT

Before we begin, we should understand the relationship between cloud computing and IoT, a line of differences between big data and cloud computing is worth mentioning. While big data refers to almost all varieties of structured or unstructured large content of data, cloud computing refers to remotely access infrastructure of services [31].

The National Institute of Standards and Technology (NIST), US Department of Commerce, explains cloud computing as an on-demand, all-pervasive, broad network model that renders remote online access to its users for a set of services like storage, infrastructure, and software applications, for achieving the economies of cost and scale. The service models of cloud computing can majorly serve in three categories, namely, Platform-as-a-Service (PaaS), Infrastructure-as-a-Service (IaaS), and even Software-as-a-Service (SaaS), while the deployment models of cloud computing can serve in four categories, namely, public cloud, community cloud, private cloud, and the hybrid cloud.

Thus, in common parlance, cloud computing comprises various computing services like networking software and analytical application besides the usual databases storage services on the remote servers through the Internet using IoT; so the operational economies of computer resources can be achieved and innovations can happen at a more flexible, optimized, and faster pace.

10.9   CONCLUSION

The introduction of IoT has been a transformative force for many firms and organizations worldwide, bringing about profound changes in the way they operate. It plays an important role in making the life much easier and simpler with least human interaction. IoT has made the work much more economical and time-managing. But IoT still presents some problems that have become a major issue while dealing the data with the IoT devices. For example, lack of security, privacy issues, and ultimately lack of trust within the devices. It also includes a requirement of a third-party server to authenticate the devices in node. It collects data from each node that can be dangerous as attackers may have the access to the central server. This chapter has presented a comprehensive review and elaborated various cases of IoT devices in various sectors and different technologies.

REFERENCES

1. Selvaraj, R., Kuthadi, V. M., & Baskar, S. (2023). Smart building energy management and monitoring system based on artificial intelligence in smart city. *Sustainable Energy Technologies and Assessments, 56*, 103090.
2. Fan, Y. J., Yin, Y. H., Da Xu, L., Zeng, Y., & Wu, F. (2014). IoT-based smart rehabilitation system. *IEEE Transactions on Industrial Informatics, 10*(2), 1568–1577.
3. Ahmed, A., Hassan, I., El-Kady, M. F., Radhi, A., Jeong, C. K., Selvaganapathy, P. R., . . . & Kaner, R. B. (2019). Integrated triboelectric nanogenerators in the era of the Internet of Things. *Advanced Science, 6*(24), 1802230.
4. Dobre, C., & Xhafa, F. (2014). Intelligent services for big data science. *Future Generation Computer Systems, 37*, 267–281.
5. Moosavi, S. R., Rahmani, A. M., Westerlund, T., Yang, G., Liljeberg, P., & Tenhunen, H. (2014, November). Pervasive health monitoring based on Internet of Things: Two case studies. In *2014 4th International Conference on Wireless Mobile Communication and Healthcare-Transforming Healthcare through Innovations in Mobile and Wireless Technologies (MOBIHEALTH)* (pp. 275–278). IEEE.
6. Asif-Ur-Rahman, M., Afsana, F., Mahmud, M., Kaiser, M. S., Ahmed, M. R., Kaiwartya, O., & James-Taylor, A. (2018). Toward a heterogeneous mist, fog, and cloud-based framework for the Internet of Healthcare Things. *IEEE Internet of Things Journal, 6*(3), 4049–4062.
7. Chirico, A., Lucidi, F., De Laurentiis, M., Milanese, C., Napoli, A., & Giordano, A. (2016). Virtual reality in health system: Beyond entertainment. A mini-review on the efficacy of VR during cancer treatment. *Journal of Cellular Physiology, 231*(2), 275–287.
8. Levine, D. M., Ouchi, K., Blanchfield, B., Diamond, K., Licurse, A., Pu, C. T., & Schnipper, J. L. (2018). Hospital-level care at home for acutely ill adults: A pilot randomized controlled trial. *Journal of General Internal Medicine, 33*(5), 729–736.
9. Lu, N., Li, T., Ren, X., & Miao, H. (2016). A deep learning scheme for motor imagery classification based on restricted Boltzmann machines. *IEEE Transactions on Neural Systems and Rehabilitation Engineering, 25*(6), 566–576.
10. Mathews, S. M., Kambhamettu, C., & Barner, K. E. (2018). A novel application of deep learning for single-lead ECG classification. *Computers in Biology and Medicine, 99*, 53–62.
11. Nadarzynski, T., Miles, O., Cowie, A., & Ridge, D. (2019). Acceptability of artificial intelligence (AI)-led chatbot services in healthcare: A mixed-methods study. *Digital Health, 5*, 2055207619871808.
12. Dargaoui, S., et al. (2024). Security issues in Internet of Medical Things. In *Blockchain and Machine Learning for IoT Security* (pp. 77–91). Chapman and Hall/CRC.
13. Dargaoui, S., Azrour, M., Allaoui, A., Guezzaz, A., Alabdulatif, A., & Alnajim, A. (2024). Internet of Things authentication protocols: Comparative study. *CMC, 79*(1), 65–91. https://doi.org/10.32604/cmc.2024.047625.
14. Bella, K., et al. (2024). An efficient intrusion detection system for IoT security using CNN decision forest. *PeerJ Computer Science, 10*, e2290. https://doi.org/10.7717/peerj-cs.2290.
15. Mohy-Eddine, M., Guezzaz, A., Benkirane, S., & Azrour, M. (2024). Malicious detection model with artificial neural network in IoT-based smart farming security. *Cluster Computing*, 1–16.
16. Hajraoui, N., Azrour, M., & Allaoui, A. E. (2023). Classification of diseases in tomato leaves with Deep Transfer Learning. *Data and Metadata, 2*, 181–181. https://doi.org/10.56294/dm2023181.

17. Mabrouki, J. et al. (2023). Geographic information system for the study of water resources in Chaâba El Hamra, Mohammedia (Morocco). In *Artificial Intelligence and Smart Environment: ICAISE'2022* (pp. 469–474). Springer.

18. Mohy-Eddine, M., Guezzaz, A., Benkirane, S., & Azrour, M. (2023). IoT-enabled smart agriculture: Security issues and applications. In *Artificial Intelligence and Smart Environment: ICAISE'2022* (pp. 566–571). Springer.

19. Hissou, H., Benkirane, S., Guezzaz, A., Azrour, M., & Beni-Hssane, A. (2023). A novel machine learning approach for solar radiation estimation. *Sustainability*, *15*(13), Art. no. 13. https://doi.org/10.3390/su151310609.

20. Presser, M., Barnaghi, P. M., Eurich, M., & Villalonga, C. (2009). The SENSEI project: Integrating the physical world with the digital world of the network of the future. *IEEE Communications Magazine*, *47*(4), 1–4.

21. Liu, J. (2016). Development and application of agricultural Internet of Things technology. *Agriculture Technology*, *36*(19), 179–180.

22. Shan, S. S. (2019). Research on the development status and countermeasures of agricultural Internet of Things. *Public Investment Guide*, *2019*(11), 220–222.

23. Gope, P., & Hwang, T. (2015). BSN-care: A secure IoT-based modern healthcare system using body sensor network. *IEEE Sensors Journal*, *16*(5), 1368–1376.

24. Borycki, E. (2019). Quality and safety in ehealth: The need to build the evidence base. *Journal of Medical Internet Research*, *21*(12), e16689.

25. Marques, G., & Pitarma, R. (2016). An indoor monitoring system for ambient assisted living based on Internet of Things architecture. *International Journal of Environmental Research and Public Health*, *13*(11), 1152. https://doi.org/10.3390/ijerph13111152.

26. Bathilde, J. B. (2018). Continuous heart rate monitoring system as an IoT edge device. In *Proceedings of 2018 IEEE Sensors Applications Symposium (SAS)* (pp. 1–6.s). Seoul.

27. Mathur, S., Kalla, A., Gür, G., Bohra, M. K., & Liyanage, M. (2023). A survey on role of blockchain for IoT: Applications and technical aspects. *Computer Networks*, *227*, 109726.

28. Huang, R., Yang, X., & Ajay, P. (2023). Consensus mechanism for software-defined blockchain in Internet of Things. *Internet of Things and Cyber-Physical Systems*, *3*, 52–60.

29. Ahaidous, K., Tabaa, M., & Hachimi, H. (2023). Towards IoT-Big Data architecture for future education. *Procedia Computer Science*, *220*, 348–355.

30. Bibri, S. E., Alexandre, A., Sharifi, A., & Krogstie, J. (2023). Environmentally sustainable smart cities and their converging AI, IoT, and big data technologies and solutions: An integrated approach to an extensive literature review. *Energy Informatics*, *6*(1), 9.

31. Nair, M. M., & Tyagi, A. K. (2023). AI, IoT, blockchain, and cloud computing: The necessity of the future. In *Distributed Computing to Blockchain* (pp. 189–206). Academic Press.

11 Smart Agro-Tech
Using IoT-Powered Intelligence and Sustainable AI Solutions to Revolutionize Crop Management

Shivani Bhardwaj, Gauarv Gupta,
Abhishek Tomar, Mukesh Tiwari,
Sultan Ahmad, and Mourade Azrour

11.1 INTRODUCTION

Half of the world's farming population still practices traditional farming, which is considered a primitive farming style. It involves utilizing natural assets, organic fertilizers, traditional tools, indigenous knowledge, and farmers' cultural beliefs [1]. Since the dawn of human civilization millions of years ago, agricultural practices have always come first. Humans adopted and used a variety of techniques and strategies to meet their basic needs, such as hunger. Traditional farming was one of the techniques they used in the past, and modern farming is one of the more recent strategies they have used. Small-scale farming is another name for traditional farming. It alludes to farming methods that are ingrained in our culture and are carried out generation after generation using antiquated methods and equipment. Along with promoting agricultural diversity and organic methods, it also entails growing crops and rearing livestock [2]. The transition of farming from traditional to smart farming is demonstrated in Figure 11.1.

11.2 TRADITIONAL FARMING

- *Combining Livestock and Crop Farming:* Integrating crops and livestock reduces risk and does not recycle resources. Crops and livestock work together in an integrated system, maximizing resource utilization through recycling. Crop residues can be used as animal feed, and livestock and by-product production can boost crop yield by increasing soil fertility and reducing the need for chemical fertilizers. While integrating crops and livestock can be beneficial, small farmers require access to knowledge, assets, and inputs to manage the system sustainably in the long run [3].

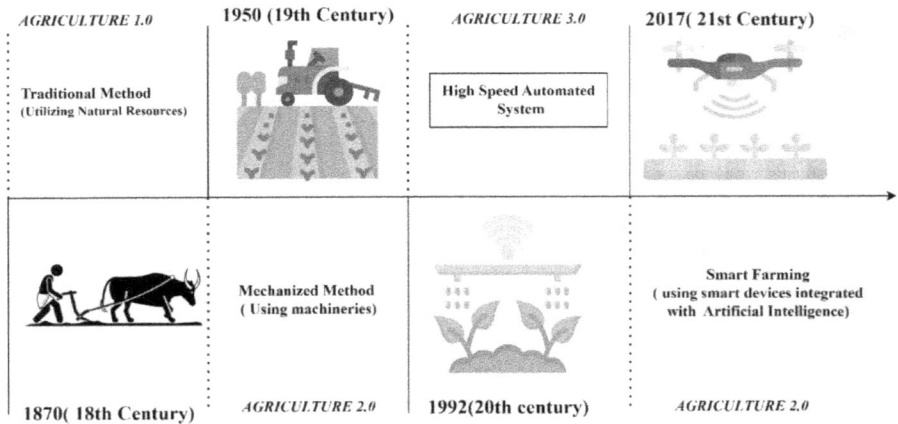

FIGURE 11.1   Transition of agriculture from traditional to modern era.

- *Water Harvesting:* Water harvesting has been a long-standing practice in drylands worldwide, leading to the development of numerous techniques. The traditional method of water harvesting is to store and collect rainwater using underground tanks and roof water storage systems. Utilization of stored water is useful for irrigation and crop production [4]. It is widely acknowledged that conserving water will increase agricultural output [5–9], particularly in arid and semiarid areas, and water harvesting is a traditional conservation technique.
- *Crop Rotation:* A traditional method of improving soil quality, reducing pests and diseases, and maximizing resource use is crop rotation, which entails growing various crops in succession on the same plot of land. The adaptability and advantages of this strategy are demonstrated by a variety of crop rotation cycles, such as three-year and four-year rotations, as well as examples involving legumes, grains, and cover crops. Conversely, planting involves cultivating two or more crops together in the same field, taking advantage of complementary interactions to boost durability, effectiveness, and resource utilization [10].
- *Shifting Cultivation:* The primary agricultural practice in the northeast Indian hills is shifting cultivation, which is also known as slash and burn agriculture. Only a small percentage of the total cultivated land, which is primarily limited to the valley lands, is under settled cultivation due to the mountainous terrain. Shifting cultivation method is a tried-and-true method of farming that is largely derived from traditional knowledge and most often developed locally [11].
- *Poly Culture:* The procedure of adding two or more species with different eating habits, ecological needs, and behaviors in the same pond to boost output is known as polyculture. Multitrophic aquaculture, coculture, and integrated aquaculture are another name for polyculture. There are three general forms of polyculture – sequential, cage-cum-pond, and direct. The principles of sustainable aquaculture align with polyculture. By maximizing

the use of available resources, it lessens the activity's negative effects on the environment, boosts producer profitability, and offers advantages related to improved ecological stability and function [12].

- *Traditional Organic Composting:* The largest contributor to the overall greenhouse gas emissions from the agriculture sector is fertilizer-driven emissions. About 75% of direct emissions from agricultural soil are caused by inorganic nitrogen fertilizers. Nitrogenous fertilizers reduce soil microbes and diversity of microbes in addition to increasing greenhouse gas emissions [13].

11.3 WHAT IS SMART AGRICULTURE?

A more efficient, less wasteful, and more productive harvest is the goal of "smart farming," which involves the integration of high-tech tools with traditional farming practices [14]. Soil management, irrigation, pest control, logistics, and crop monitoring all are improved by these technological advancements. Unmanned aerial vehicles, video surveillance systems, GPS, agricultural information management platforms [15–18], sensors that measure soil composition, light intensity, humidity, and temperature are just a few of the technological tools used by smart farms. Improved accuracy and long-term viability in farming are outcomes of the real-time data collection and decision-making made possible by these technologies [19].

11.3.1 TYPES OF SMART FARMING

In the current era, there are several methods of smart farming that will help to ensure the sustainability of farming. Following diverse farming methods can result in precise crop output and production. It can also help to reduce hard effort in the field and boost smart work for greater productivity (Table 11.1).

TABLE 11.1
Types of Smart Farming Using Targeted Technologies and Their Benefits

S. No./ References	Types of Smart Farming	Targeted Technology Used	Applications	Benefits
[44]	Precision farming	Sensors for humidity, soil, and GPS for remote access	Monitoring crop health, pest detection, and appropriate fertilization	More production by utilizing small land while using a smaller number of resources
[45]	Hydroponic	Soil-less cultivation technology	To produce fresh and healthy vegetables and fruits	Produce crops in the desert without soil and in fertilized land
[46]	Greenhouse farming	ML algorithms and IoT sensors	Provide accurate analysis and a healthy environment for crops	Increase the productivity and reduce the cost

(Continued)

TABLE 11.1   (Continued)
Types of Smart Farming Using Targeted Technologies and Their Benefits

S. No./ References	Types of Smart Farming	Targeted Technology Used	Applications	Benefits
[47]	Aquaculture	Arduino, sensors, Raspberry pi, and fuzzy method	Automatic aqua farm, monitoring water quality of the farm using mobile devises	Provide certain temperatures to increase the survival rate of aquatics
[48]	Agro-Robotics	Usage of motors, Bluetooth, and Arduino	Used for automated sowing, fertilizing, and irrigation systems	Reduce human interference and use precise farming
[49]	Smart irrigation system	Soil, humidity, nutrients, and other sensors are used Mobile application for automated controlling	Drip irrigation, sprinkle irrigation	To avoid wastage of water and provide water to a barren land
[50]	Vertical farming	Sensors for different things to maintain environmental conditions	Indoor farming	To set light, humidity, waterflow according to the need of a crop

11.4   AI AND IoT EMERGING TECHNOLOGIES IN SUSTAINABLE AGRICULTURE

Agricultural use of modern innovations such as artificial intelligence, Internet of Things, and machine learning boosts productivity while also boosting sustainability and decision-making processes. To lowering risks, improving sustainability, and providing growers with the opportunity to make more informed decisions, traditional agricultural processes can be improved through the implementation of scalable technology solutions and automation [20]. The combination of artificial intelligence and the Internet of Things presents a glimmer of hope by making it possible to farm more efficiently using real-time monitoring, data analysis, and administration of agricultural activities. The implementation of these technologies in many aspects of farming, such as precision agriculture, crop yield forecasts, and the management of resources, is investigated in depth throughout this chapter. Specifically, it investigates how the analytical capabilities of artificial intelligence

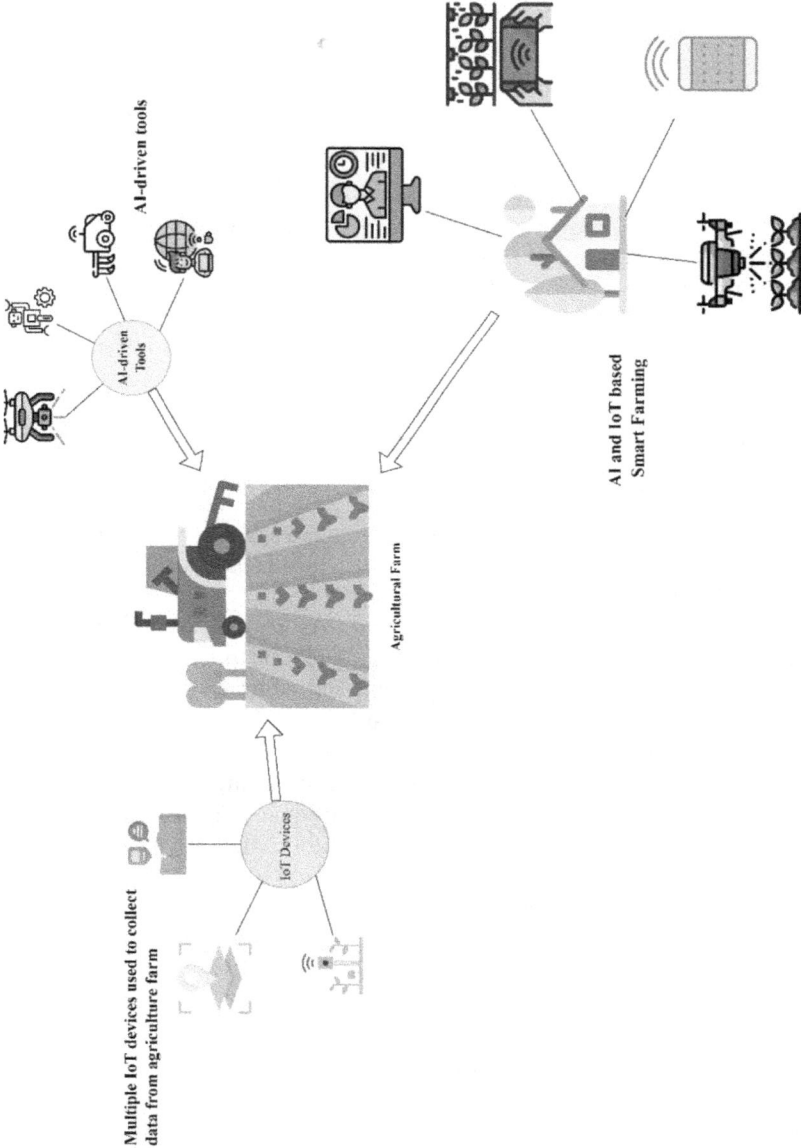

FIGURE 11.2 AI- and IoT-driven intelligent system for sustainable farming.11.2.3

can be used to forecast and treat crop illnesses, automate agricultural activities, and provide assistance with decision-making, while the Internet of Things' network of sensors and devices makes it possible to track and monitor farm conditions in real time [21]. New technologies like cloud computing, automation systems, and artificial intelligence are revolutionizing how we combine the digital and physical worlds in the era of Industry 4.0 [22]. The emerging of the technology for sustainable agriculture is shown in Figure 11.2.

11.4.1 Challenges Highlighted Achieving Sustainability within Smart Farming

- *High Initial Costs:* Individual farms face significant barriers due to the substantial expenses of technology execution, especially within nations with poor infrastructure where a lack of knowledge and skills exacerbate the issue. As a result, large-industrialized farms usually have access to the most recent advancements, putting smaller agricultural business at a competitive disadvantage. This variability may inhibit overall progress in modern and sustainable farming in some locations [23].
- *Technology Complexity:* Technological difficulties offer an enormous threat to achieving the environmental benefits of sustainable farming. They either impede implementation or cause ambiguity, making farmers unwilling to employ SF technologies. Based on our literature review, we identified three major types of technical barriers – data protection and security, standardization, and infrastructure needs [24].
- *Data Privacy and Security Concerns:* Farmers who use smart infrastructure worry that their data could be made public or pillaged by competitors. Data security is therefore essential, and contracts with technology suppliers should have certain provisions. Even if many smart farming methods handle non-personal data, there are significant issues when this data is connected to specific personally identifiable information (PII). For instance, animal data that directly refers to the owner; agricultural circumstances that are connected to the personal information of farmers. In this situation, privacy rules ought to be incorporated to forbid the processing of personal data to a certain extent. When smart farming devices, like tractors or drones, have the ability to monitor their users, further problems could occur [25].
- *Dependence on Conductivity:* The need on dependable connection is a significant barrier to attaining sustainability in smart farming. Intelligent agricultural equipment, including IoT devices, sensors, and automated machinery, necessitates continuous Internet access to transmit data and facilitate real-time decision-making. Rural areas, where several farms are located, often lack reliable, high-speed Internet connections, thereby constraining the efficacy of these technologies. This connectivity gap may lead to delay essential farming operations [26].

11.5   PRECISION FARMING

Precision farming is known as smart farming or smart agriculture; it has become a game-changing way to tackle the intricate problems that contemporary agriculture faces. Precision farming is crucial because it enables a sustainable agriculture future [27]. By reducing environmental loading by applying fertilizers and pesticides just where they are needed, when they are needed, PA can assist in many ways to the long-term sustainability of production agriculture. This confirms the intuitive concept. Reduced input losses owing to nutrient imbalances, weed escapes, insect damage, and so on, as well as losses from excess applications, are some of the environmental benefits of precision agriculture. Lessening the emergence of pesticide resistance is another advantage [28].

11.5.1   KEY COMPONENTS OF PRECISION FARMING

Technologies and methods that maximize agricultural inputs and procedures based on data-driven insights are essential elements of precision farming. The primary elements are as follows with a diagram representation in Figure 11.3.

- *Geographic Information Systems:* This technology offers a user-friendly interface that facilitates an integrated approach to handling agricultural information. Using Geographic Information Systems, farm information can be used for integrated management. Plot-by-plot queries regarding crop varieties planted, fertilizers used, pest infestations, and yield are made possible by an intuitive interface [29].
- *Remote Sensing:* In agriculture, remote sensing makes use of satellites or unmanned aerial vehicles to track and manage crop health, soil conditions, and overall output. Remote sensing can identify oscillations in agricultural stress, soil moisture levels of difficulty, vegetation vitality, and infestations of bugs by collecting data across several wavelengths. Employing this

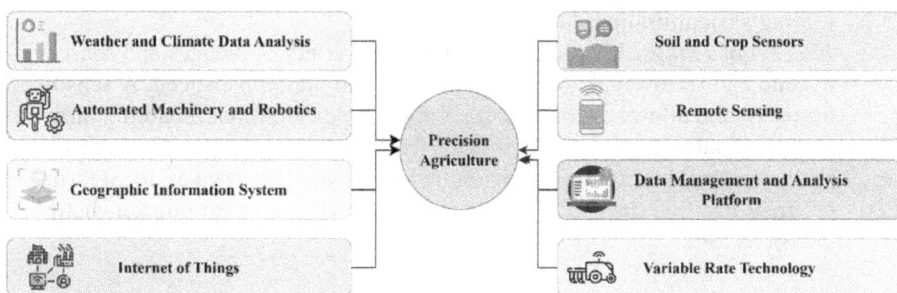

FIGURE 11.3   The key components of precision farming.

technology, farmers may assess crop growth, predict yields, give precise treatments, enhance resource efficiency, and save expenses [30].

- *Soil and Crop Sensors:* Soil and crop sensors serve as vital devices in intelligent farming, offering precise monitoring and management of soil health and crop status. Devices gather real-time data on soil characteristics like moisture, pH, temperature, nutrient composition, and salt. This information helps farmers to improve the yield of crops [31].
- *Variable Rate Technology (VRT):* Numerous firms are currently developing controllers, sprayers, air spreaders, anhydrous ammonia systems, and herbicide applicators for variable rate technology applications. An optical sensor to quantify flow rates of granular fertilizer within airstreams for the feedback regulation of a variable-rate spreader has been developed [32].
- *Weather and Climate Data Analysis:* Weather and climatic data are important across the crop cycle, from choosing the best crop or variety to post-harvest activities and marketing. If provided ahead of time, the data can encourage farmers to mobilize and activate their own resources so that they can benefit from the prudent use of expensive inputs. Therefore, weather data can assist the farmer in maximizing the use of natural resources to increase agricultural output, in terms of both quantity and quality. The analyses also provide meteorological information in conjunction with weather-sensitive management [33].
- *Data Management and Analysis Platforms:* Good data management systems compile and examine information from several sources, including sensors, drones, and satellite photos, to give farmers useful information. Cloud-based technologies are frequently used in many industries and businesses, and agriculture is one of them. The way they analyze the data appears to be effective and well-mannered, but it wasn't until stakeholders began to realize its potential advantages in agriculture. Some of the popular agriculture companies claim that crop revenues could rise by over US$20 billion annually if farmers applied this technology to their farming [34].
- *Automated Machinery and Robotics:* By implementing robots, farming in these areas has the potential to leap directly to sophisticated automation. This may lead to more productive, less labor-intensive production and sustaining agricultural productivity [35].
- *Internet of Things:* IoT permits an extensive variety of devices and sensors to send and receive data over the Internet at an incredible speed. A sensor-based system allows continuous tracking of field conditions, enabling farmers to evaluate the status of their fields from any location remotely. Together with information from third parties, including weather services, local farmers may use this data to improve data analysis and make quicker, better decisions [36].

11.6 IoT-DRIVEN INTELLIGENCE IN CROP MANAGEMENT

Achieving sustainable agricultural systems in crop production requires a balance of ecological, economic, and social elements. Despite the growing effect of labor, market, and policy forces on farming, ecological control is still a top priority, necessitating long-term choices by farmers based on climate, landscape, social needs, resources, and market demands [37]. Over the last two decades, the Internet of Things (IoT) has grown into a unified reality made up of a collection network of devices connected in a dynamic (and typically asynchronous) environment. This allows for the provision of vast amounts of data to feed machine learning algorithms and can also respond proactively to environmental stimuli using actuators designed to reduce the need for human intervention. Because it enables farmers to improve ecological control through accurate monitoring and response systems, this variability emphasizes the rising significance of IoT-driven crop management. This helps farmers to maximize output efficiency even in less controlled situations [38].

As shown in Table 11.2, IoT-driven solutions are significantly enhancing crop management by providing precise monitoring and automated control across a few agricultural phases, from supply chain management and irrigation to soil health and crop disease detection. With the help of these tools, farmers can make data-driven decisions that boost sustainability and efficiency by boosting resistance to environmental changes, preserving resources, and increasing agricultural output. Still,

TABLE 11.2
IoT-Driven Intelligent Application for Sustainable Farming

References	IoT-Driven Application	Functionalities	Benefits	Challenges
[51]	Soil health monitoring	Used to measure the soil nutrient NPK, pH, moisture, and temperature	This may lead to increases in crop production and precise farming	Complexity in data integrity and high initial costs
[52]	Crop health monitoring	Help farmers to detect crop disease, pest infection, and some physical parameters such as humidity, moisture, and fertilizers		Data security and data analysis complexity
[53]	Precision irrigation	Automatic irrigation system as per the requirements of daily basis environmental conditions	Increases the effectiveness of crop hydration while using less water	Reliable connectivity and infrastructure increase the effectiveness of crop hydration while using less water

(Continued)

TABLE 11.2 (Continued)
IOT-Driven Intelligent Application for Sustainable Farming

References	IoT-Driven Application	Functionalities	Benefits	Challenges
[54]	Climate adaptive system	Makes use of meteorological data to adjust the timing of agricultural activities, including cultivation, growing, and reaping	Improves the susceptibility of crops to variations in the climate	Optimize the resilience of crops to the effects of fluctuations in the climate
[55]	Pest and disease control	Smart pest and disease management system using multiple sensors	Reduce input costs	Integrating with real-time data
[56]	Supply chain monitoring	Monitors temperature and humidity levels of harvested crops across the supply chain	Minimizes waste and improves logistics	Issues with data sharing and privacy

issues including data integrity, privacy concerns, high upfront costs, and the need for dependable infrastructure can occasionally make it challenging to successfully adopt these solutions. These barriers need to be eliminated if IoT is to reach its full potential in sustainable agriculture.

11.7 CLIMATE RESILIENCE THROUGH SMART FARMING

Increases in extreme weather have led to several issues, drastically lowered crop yields, and prevented agricultural cultivation in the context of climate change. Crop rotation is a significant technique that may be used to improve the climate resilience of the agricultural production system and efficiently solve the drawbacks of the continuous crop technology that is now in use. The rotation of crops is an essential component of numerous national initiatives, such as those pertaining to food security, the development of ecological environments, and the renovation of rural areas [39].

11.7.1 EFFECTS OF CLIMATE CHANGE ON SUSTAINABLE AGRICULTURE

- *Rising Temperature:* The increase in temperature causes uneven crop growth, which can lead to low production and higher market pricing. Increased average temperatures lead to a reduced amount of soil organic matter and result in heightened potential for soil erosion with increased rates of flow of water and organic and inorganic compounds [40].
- *Unpredictable Precipitation:* Uneven periodic distribution of rainfall and rising temperatures are important problems in the area. Meteorological records show that the mean maximum temperature has steadily increased, affecting natural systems by changing phenological processes in plants, enhancing evapotranspiration rates, and raising agricultural pest and disease outbreak risks [41].
- *Pests and Disease:* Agricultural parasites and crop production are significantly affected by climate change and extreme weather events. It is anticipated that global climate warming will result in an expansion of the geographic range of insects, an increase in overwintering survival of several generations, a higher risk of invasive insect species and insect-transmitted plant diseases, and changes in their interaction with natural enemies and host plants, as temperature is the most significant environmental factor influencing insect population dynamics [42].
- *Extreme Weather Conditions:* Agriculture is greatly affected by climate abnormalities and extreme weather occurrences. Natural disasters like drought, flash floods, unexpected rainfall, frost, hail, and storms account for a large portion of the global agricultural industry's annual crop losses. Reduce the impact of weather events and climatic anomalies on production levels, land resources, and other assets like structures and infrastructure as well as natural ecosystems that are vital to agricultural operations with high preparedness, prior knowledge of the timing and magnitude of these events, and effective recovery plans [43].

11.8 REAL-WORLD APPLICATIONS AND CASE STUDIES

The following real-world case studies highlight several applications that have effectively integrated technology to address environmental concerns, optimize resources, and increase production to demonstrate the influence of IoT and data-driven strategies in sustainable agriculture.

Table 11.3 demonstrates how IoT is promoting efficient and sustainable farming methods. IoT solutions allow organic production in vertical farming by automatically controlling fertilizer and irrigation. By improving water quality and incorporating methods from various geographical areas, IoT facilitates sustainable irrigation for rice paddy management. IoT benefits dairy farmers by improving resource management, which increases milk production and reproduction rates. Through the analysis of agricultural data and the prevention of crop losses, IoT is utilized in precision pest control

TABLE 11.3
Real-World Applications and Case Studies

References	Applications Area	Description	Case Study
[22]	Wangree Health Factory Company	Achieving year-round organic production, automated system manages watering and nutrient levels	Wangree factory combines vertical farming with IoT to achieve organic production
[57]	Rice Paddy Management	Maintain good water quality, sustainable irrigation practice for rice management	This study compares the Subak irrigation system in Indonesia and the Tameike system in Japan
[58]	Dairy Farm Monitoring	Increases milk yield and reproduction rates, managing the resources and monitoring	Improving the sustainability and efficiency in dairy farming
[59]	Precision Pest Control	Prevents pest infections, crop management, prevents economic losses	Analyzes the yield data from three farms in Ontario
[60]	Urban Vertical Farming	Significant environmental effect decreases when coir is used as the growth medium instead of gardening soil	Understanding the impact of environment using vertical hydrophobic farm in Stockholm, Sweden

to maximize pest management. The significant decreases in environmental effects that result from using coir as a growing medium instead of dirt in urban vertical farming demonstrates the promise of IoT for environment-friendly urban agriculture.

11.7 CONCLUSION AND FUTURE TRENDS

In conclusion, sustainable development is essential to addressing the issues of rapid expansion, climate change, and food security. Information and communication technology integration has transformed agriculture to increase precision, production, and environmental sustainability. Farmers may increase agricultural yields, reduce waste, and maximize resource use by assisting this technology. By 2050, food consumption is expected to climb by 70%, while ecological balance is maintained. Still the significant challenges persist, including high cost of implementation, data privacy concerns, and the need for dependable connectivity, particularly for smaller farms in developing regions. Overcoming these barriers requires coordinated efforts in policy support and infrastructure investment and the development of user-friendly technologies. Enhancing AI-driven IoT and precision resources management may lead toward the sustainable development goal, ensuring food quality and security for future generations.

REFERENCES

[1] Z. Mushtaq, H. Mushtaq, S. Faizan, and M. A. Parray, "Microbial Degradation of Organic Constituents for Sustainable Development," *Microbiota and Biofertilizers*, vol. 2, 2021. https://doi.org/10.1007/978-3-030-61010-4_5.

[2] L. S. Saini, "Nanotechnology for Pest Management: A Promising Future," *Agri Roots*, vol. 1, no. 4, pp. 5–8, 2023.

[3] V. Gupta, P. K. Rai, and K. S. Risam, "Integrated Crop-Livestock Farming Systems: A Strategy for Resource Conservation and Environmental Sustainability," *Indian Research Journal of Extension Education*, vol. II, pp. 49–54, 2012.

[4] T. Oweis, D. Prinz, and A. Hachum, "Water Harvesting Indigenous Knowledge for the Future of the Drier Environments," *Journal of Chemical Information and Modeling*, vol. 53, no. 9, pp. 1689–1699, 2013. https://doi.org/10.1017/CBO9781107415324.004.

[5] M. Azrour, et al., "A Survey of Machine and Deep Learning Applications in the Assessment of Water Quality," in *World Sustainability Series*, vol. Part F2854, 2024, pp. 471–483. https://doi.org/10.1007/978-3-031-56292-1_38.

[6] S. Nasrdine, M. Benchrifa, N. Ben-Lhachemi, J. Mabrouki, M. Slaoui, and M. Azrour, "New Design of an Inclined Solar Distiller for Freshwater Production: Experimental Study," in *World Sustainability Series*, vol. Part F2854, 2024, pp. 447–453. https://doi.org/10.1007/978-3-031-56292-1_36.

[7] M. Azrour, J. Mabrouki, A. Guezzaz, S. Benkirane, and H. Asri, "Implementation of Real-Time Water Quality Monitoring Based on Java and Internet of Things," in *Integrating Blockchain and Artificial Intelligence for Industry 4.0 Innovations*, S. Goundar and R. Anandan, Eds., in EAI/Springer Innovations in Communication and Computing, Cham: Springer International Publishing, 2024, pp. 133–143. https://doi.org/10.1007/978-3-031-35751-0_8.

[8] J. Mabrouki, et al., "Geographic Information System for the Study of Water Resources in Chaâba El Hamra, Mohammedia (Morocco)," in *Artificial Intelligence and Smart Environment: ICAISE'2022*, Springer, 2023, pp. 469–474.

[9] M. Mohy-Eddine, M. Azrour, J. Mabrouki, F. Amounas, A. Guezzaz, and S. Benkirane, "Embedded Web Server Implementation for Real-Time Water Monitoring," in *Advanced Technology for Smart Environment and Energy*, J. Mabrouki, A. Mourade, A. Irshad, and S. A. Chaudhry, Eds., in Environmental Science and Engineering, Cham: Springer International Publishing, 2023, pp. 301–311. https://doi.org/10.1007/978-3-031-25662-2_24.

[10] S. Pandey, "Crop Rotation and Intercropping Techniques," no. April, pp. 38–54, 2024. https://doi.org/10.5281/zenodo.11051201.

[11] shifting_cultivation.pdf.

[12] M. Wang and M. Lu, "Tilapia Polyculture: A Global Review," *Aquaculture Research*, vol. 47, no. 8, pp. 2363–2374, 2016. https://doi.org/10.1111/are.12708.

[13] R. Singh and G. S. Singh, "Traditional Agriculture: A Climate-Smart Approach for Sustainable Food Production," *Energy, Ecology and Environment*, vol. 2, no. 5, pp. 296–316, 2017. https://doi.org/10.1007/s40974-017-0074-7.

[14] A. Muniasamy, "Machine Learning for Smart Farming: A Focus on Desert Agriculture," in *2020 International Conference on Computing and Information Technology, ICCIT 2020*, 2020, pp. 438–442. https://doi.org/10.1109/ICCIT-144147971.2020.9213759.

[15] S. Dargaoui, et al., "IoT-Driven Smart Agriculture: Security Issues and Authentication Schemes Classification," in *Proceeding of the International Conference on Connected Objects and Artificial Intelligence (COCIA2024)*, Y. Mejdoub and A. Elamri, Eds.,

Cham: Springer Nature Switzerland, 2024, pp. 61–66. https://doi.org/10.1007/978-3-031-70411-6_10.

[16] S. Dargaoui, et al., "Internet-of-Things-Enabled Smart Agriculture: Security Enhancement Approaches," in *2024 4th International Conference on Innovative Research in Applied Science, Engineering and Technology (IRASET)*, May 2024, pp. 1–5. https://doi.org/10.1109/IRASET60544.2024.10548705.

[17] N. Ben-Lhachemi, M. Benchrifa, S. Nasrdine, J. Mabrouki, M. Slaoui, and M. Ade Azrour, "Effect of IoT Integration in Agricultural Greenhouses," in *Technical and Technological Solutions towards a Sustainable Society and Circular Economy*, J. Mabrouki and A. Mourade, Eds., Cham: Springer Nature Switzerland, 2024, pp. 435–445. https://doi.org/10.1007/978-3-031-56292-1_35.

[18] K. El-Moustaqim, J. Mabrouki, M. Azrour, M. Hadine, and D. Hmouni, "Enabling Smart Agriculture Through Integrating the Internet of Things in Microalgae Farming for Sustainability," in *Smart Internet of Things for Environment and Healthcare*, M. Azrour, J. Mabrouki, A. Alabdulatif, A. Guezzaz, and F. Amounas, Eds., Cham: Springer Nature Switzerland, 2024, pp. 209–222. https://doi.org/10.1007/978-3-031-70102-3_15.

[19] E. Navarro, N. Costa, and A. Pereira, "The Development of Administration Theories," *Sensors (Switzerland)*, vol. 20, no. 15, pp. 1–29, 2020.

[20] V. Balaska, Z. Adamidou, Z. Vryzas, and A. Gasteratos, "Sustainable Crop Protection via Robotics and Artificial Intelligence Solutions," *Machines*, vol. 11, no. 8, pp. 1–15, 2023. https://doi.org/10.3390/machines11080774.

[21] A. H. Abdul Hussein, K. A. Jabbar, A. Mohammed, and H. M. Al-Jawahry, "AI and IoT in Farming: A Sustainable Approach," *E3S Web of Conferences*, vol. 491, pp. 1–10, 2024. https://doi.org/10.1051/e3sconf/202449101020.

[22] S. Santiteerakul, A. Sopadang, and K. Y. Tippayawong, "The Role of Smart Technology in Sustainable Agriculture: A Case Study of Wangree Plant Factory," *Sustainability*, vol. 12, no. 11, p. 4640, 2020.

[23] A. Walter, R. Finger, R. Huber, and N. Buchmann, "Smart Farming Is Key to Developing Sustainable Agriculture," vol. 114, no. 24, pp. 6148–6150, 2017. https://doi.org/10.1073/pnas.1707462114.

[24] S. Lieder and C. Schröter-Schlaack, "Smart Farming Technologies in Arable Farming: Towards a Holistic Assessment of Opportunities and Risks," *Sustainability*, vol. 13, no. 12, p. 6783, 2021.

[25] M. Gupta, M. Abdelsalam, and S. Mittal, "Security and Privacy in Smart Farming: Challenges and Opportunities," *IEEE Access*, vol. 8, pp. 34564–34584, 2020. https://doi.org/10.1109/ACCESS.2020.2975142.

[26] F. Boudet, G. K. Macdonald, B. E. Robinson, and L. H. Samberg, "Rural-Urban Connectivity and Agricultural Land Management Across the Global South," *Global Environmental Change*, vol. 60, no. September 2019, p. 101982, 2020. https://doi.org/10.1016/j.gloenvcha.2019.101982.

[27] S. Biswas, S. Halder, B. Koley, E. Adak, and S. Sengupta, "The Role of Precision Farming in Sustainable Agriculture: An Overview," *International Journal of Agriculture Extension and Social Development*, vol. 7, no. 4, pp. 219–228, 2024. https://doi.org/10.33545/26180723.2024.v7.i4c.536.

[28] R. Bongiovanni and J. Lowenberg-Deboer, "Precision Agriculture and Sustainability," *Precision Agriculture*, vol. 5, no. 4, pp. 359–387, 2004. https://doi.org/10.1023/B:PRAG.0000040806.39604.aa.

[29] P. D. Sreekanth, K. V. Kumar, S. K. Soamand, and C. H. Srinivasarao, "Spatial Decision Support Systems for Smart Farming Using Geo-Spatial Technologies," in *National Conference on Application of Geospatial Technologies and IT in Smart Farming*. Dharwad: University of Agricultural Science, Dharwad, 2018, pp. 118–122.

[30] Y. Inoue, "Satellite- and Drone-Based Remote Sensing of Crops and Soils for Smart Farming–A Review," *Soil Science and Plant Nutrition*, vol. 66, no. 6, pp. 798–810, 2020. https://doi.org/10.1080/00380768.2020.1738899.

[31] J. V. Ganesh, et al., "Smart Farming System Using IoT for Efficient Crop Growth," in *2024 3rd International Conference for Advancement in Technology (ICONAT)*. IEEE, 2024, pp. 1–5.

[32] N. Zhang, M. Wang, and N. Wang, "Precision Agriculture – A Worldwide Overview," *Computers and Electronics in Agriculture*, vol. 36, pp. 113–132, 2002.

[33] M. Lecture, "Weather Information for Sustainable Agriculture in India," *Journal of Agricultural Physics*, vol. 13, no. 2, pp. 89–105, 2013.

[34] A. Kamilaris, A. Kartakoullis, and F. X. Prenafeta-boldú, "A Review on the Practice of Big Data Analysis in Agriculture," *Computers and Electronics in Agriculture*, vol. 143, no. September, pp. 23–37, 2017. https://doi.org/10.1016/j.compag.2017.09.037.

[35] J. Lowenberg-DeBoer, I. Y. Huang, V. Grigoriadis, and S. Blackmore, "Economics of Robots and Automation in Field Crop Production," *Precision Agriculture*, vol. 21, no. 2, pp. 278–299, 2020. https://doi.org/10.1007/s11119-019-09667-5.

[36] V. Porkodi, D. Yuvaraj, A. S. Mohammed, M. Sivaram, and V. Manikandan, "IoT in Agriculture," *Journal of Advanced Research in Dynamical and Control Systems*, vol. 10, no. 14, pp. 1986–1991, 2018. https://doi.org/10.48175/ijarsct-1351.

[37] L. O. Colombo-Mendoza, M. A. Paredes-Valverde, M. D. P. Salas-Zárate, and R. Valencia-García, "Internet of Things-Driven Data Mining for Smart Crop Production Prediction in the Peasant Farming Domain," *Applied Sciences (Switzerland)*, vol. 12, no. 4, 2022. https://doi.org/10.3390/app12041940.

[38] G. Vitali, M. Francia, M. Golfarelli, and M. Canavari, "Crop Management with the IoT: An Interdisciplinary Survey," *Agronomy*, vol. 11, no. 1, pp. 1–18, 2021. https://doi.org/10.3390/agronomy11010181.

[39] T. Yu, L. Mahe, Y. Li, X. Wei, X. Deng, and D. Zhang, "Benefits of Crop Rotation on Climate Resilience and Its Prospects in China," *Agronomy*, vol. 12, no. 2, pp. 1–18, 2022. https://doi.org/10.3390/agronomy12020436.

[40] G. S. Malhi, M. Kaur, and P. Kaushik, "Impact of Climate Change on Agriculture and Its Mitigation Strategies: A Review," *Sustainability (Switzerland)*, vol. 13, no. 3, pp. 1–21, 2021. https://doi.org/10.3390/su13031318.

[41] R. Y. M. Kangalawe, C. G. Mung'ong'o, A. G. Mwakaje, E. Kalumanga, and P. Z. Yanda, "Climate Change and Variability Impacts on Agricultural Production and Livelihood Systems in Western Tanzania," *Climate and Development*, vol. 9, no. 3, pp. 202–216, 2017. https://doi.org/10.1080/17565529.2016.1146119.

[42] S. Skendžić, M. Zovko, I. P. Živković, V. Lešić, and D. Lemić, "The Impact of Climate Change on Agricultural Insect Pests," vol. 12, no. 5, 2021. https://doi.org/10.3390/insects12050440.

[43] V. Radović, et al., "Extreme Weather and Climatic Events on Agriculture as a Risk of Sustainable Development," *Економика пољопривреде*, vol. 62, no. 1, pp. 181–191, 2015.

[44] G. Gyarmati and T. Mizik, "The Present and Future of the Precision Agriculture," in *SOSE 2020 – IEEE 15th International Conference of System of Systems Engineering, Proceedings*, pp. 593–596, 2020. https://doi.org/10.1109/SoSE50414.2020.9130481.

[45] S. Khan, A. Purohit, and N. Vadsaria, "Hydroponics: Current and Future State of the Art in Farming," *Journal of Plant Nutrition*, vol. 44, no. 10, pp. 1515–1538, 2020. https://doi.org/10.1080/01904167.2020.1860217.

[46] P. Dedeepya, U. S. A. Srinija, M. Gowtham Krishna, G. Sindhusha, and T. Gnanesh, "Smart Greenhouse Farming Based on IOT," in *Proceedings of the 2nd International*

Conference on Electronics, Communication and Aerospace Technology, ICECA 2018, no. Iceca, pp. 1890–1893, 2018. https://doi.org/10.1109/ICECA.2018.8474713.

[47] K. L. Tsai, L. W. Chen, L. J. Yang, H. Shiu, and H. W. Chen, "IoT Based Smart Aquaculture System with Automatic Aerating and Water Quality Monitoring," *Journal of Internet Technology,* vol. 23, no. 1, pp. 177–184, 2022. https://doi.org/10.53106/1607 92642022012301018.

[48] D. D. Patil, et al., "IOT Sensor-Based Smart Agriculture Using Agro-Robot," in *IoT Based Smart Applications,* Cham: Springer International Publishing, 2022, pp. 345–361.

[49] D. Bhavsar, B. Limbasia, Y. Mori, M. Imtiyazali Aglodiya, and M. Shah, "A Comprehensive and Systematic Study in Smart Drip and Sprinkler Irrigation Systems," *Smart Agricultural Technology,* vol. 5, no. April, p. 100303, 2023. https://doi.org/10.1016/j.atech.2023.100303.

[50] E. Kaiser, et al., "Vertical Farming Goes Dynamic: Optimizing Resource Use Efficiency, Product Quality, and Energy Costs (In press)," *Frontiers in Plant Sciences,* no. September, 2024. https://doi.org/10.3389/fsci.2024.1411259.

[51] D. K. Sreekantha and A. M. Kavya, "Agricultural Crop Monitoring Using IOT – A Study," in *Proceedings of 2017 11th International Conference on Intelligent Systems and Control, ISCO 2017,* pp. 134–139, 2017. https://doi.org/10.1109/ISCO.2017.7855968.

[52] K. Tyagi, A. Karmarkar, S. Kaur, S. Kulkarni, and R. Das, "Crop Health Monitoring System," in *2020 International Conference for Emerging Technology, INCET 2020,* pp. 5–9, 2020. https://doi.org/10.1109/INCET49848.2020.9154110.

[53] O. Adeyemi, I. Grove, S. Peets, and T. Norton, "Advanced Monitoring and Management Systems for Improving Sustainability in Precision Irrigation," *Sustainability (Switzerland),* vol. 9, no. 3, pp. 1–29, 2017. https://doi.org/10.3390/su9030353.

[54] "Agriculture in a Changing Climate Hanging Environmental Constraints Facing," pp. 1–10. https://doi.org/10.1007/978-3-030-15519-3.

[55] S. M. Haldhar, G. C. Jat, H. L. Deshwal, J. S. Gora, and D. Singh, "Insect Pest and Disease Management in Organic Farming," in *Towards Organic Agriculture,* B. Gangwar and N. K. Jat, Eds., New Delhi: Today & Tomorrow's Printers and Publishers, 2017, pp. 359–390.

[56] I. C. Somashekhar, J. Raju, and Hemapatil, "Agriculture Supply Chain Management: A Scenario in India," *TIJRP RJSSM Research Journal of Social Science and Management,* vol. 4, no. 7, pp. 89–99, 2014.

[57] M. S. Jansing, F. Mahichi, and R. Dasanayake, "Sustainable Irrigation Management in Paddy Rice Agriculture: A Comparative Case Study of Karangasem, Indonesia and Kunisaki, Japan," *Sustainability (Switzerland),* vol. 12, no. 3, 2020. https://doi.org/10.3390/su12031180.

[58] M. Bovo, et al., "A Smart Monitoring System for a Future Smarter Dairy Farming," in *2020 IEEE International Workshop on Metrology for Agriculture and Forestry, MetroAgriFor 2020 – Proceedings,* pp. 165–169, 2020. https://doi.org/10.1109/MetroAgriFor50201.2020.9277547.

[59] C. Liang and T. Shah, "IoT in Agriculture: The Future of Precision Monitoring and Data-Driven Farming," *Eigenpub Review of Science and Technology,* vol. 7, no. 1, pp. 85–104, 2023.

[60] V. Capmourteres, J. Adams, A. Berg, E. Fraser, C. Swanton, and M. Anand, "Precision Conservation Meets Precision Agriculture: A Case Study from Southern Ontario," *Agricultural Systems,* vol. 167, no. May, pp. 176–185, 2018. https://doi.org/10.1016/j.agsy.2018.09.011.

12 Introduction to Smart Agriculture Using IoT

K.A. Vinodhini, T. Anstey Vathani,
K.A. Varun Kumar, M.J. Carmel Mary
Belinda, and Mourade Azrour

12.1 INTRODUCTION

Smart agriculture represents a holistic approach to farming that leverages technology to optimize production, enhance sustainability, and improve resilience. This chapter explores the multifaceted challenges in modern agriculture and the diverse roles that IoT and AI play in addressing these challenges [1]. Ensuring food quality and safety throughout the supply chain is becoming increasingly complex as food systems become more globalized. Contamination, adulteration, and foodborne illnesses pose significant risks to public health and the food industry. Modern agricultural practices have contributed to significant biodiversity loss, threatening ecosystem stability and resilience. Maintaining biodiversity is crucial for pest control, pollination, and overall ecosystem health [2]. Agriculture is an energy-intensive sector, relying heavily on fossil fuels for machinery, transportation, and the production of inputs like fertilizers. Improving energy efficiency and transitioning to renewable energy sources is a growing challenge. Small-scale farmers, especially in developing countries, often struggle with limited access to markets and up-to-date information on prices, weather, and best practices. Smart agriculture represents a paradigm shift in farming practices, leveraging cutting-edge technologies to address the complex challenges facing the global food system. This approach integrates advanced data collection, analysis, and automation to optimize agricultural processes, enhance productivity, and promote sustainability.

In this chapter, we consider several key aspects of smart farming, beginning with an exploration of the history and evolution of smart farming practices, drawing links between early methods and contemporary innovations. We also examine the many challenges facing modern agriculture, including climate change, resource scarcity, and food security, which require the adoption of smarter farming techniques. In addition, we analyze the crucial roles of the Internet of Things (IoT) and artificial intelligence (AI) in revolutionizing agricultural operations, enhancing productivity, and ensuring sustainability. We provide a clear definition and concept of smart agriculture, highlighting its fundamental principles and objectives. Furthermore, we discuss how IoT enables the transformation of agriculture through real-time monitoring and data-driven decision-making, while addressing the importance of data

DOI: 10.1201/9781003527664-12

collection and management for optimized agricultural outcomes. The chapter will also cover the transformative role of AI in agriculture, discussing its applications in crop management, predictive analytics, and operational efficiency. Finally, we'll explore advances in agricultural robotics and automation, examining how these technologies are contributing to labor efficiency, precision farming, and, ultimately, the future of food production. This comprehensive analysis aims to provide readers with an in-depth understanding of information and communication technologies.

12.2   DEFINITION AND CONCEPT OF SMART AGRICULTURE

Smart Agriculture, also known as Precision Agriculture or Digital Farming, represents a revolutionary approach to farming that integrates cutting-edge technologies, data analytics, and interconnected systems to optimize agricultural processes. This innovative paradigm aims to enhance productivity, improve sustainability, and increase the resilience of farming operations in the face of modern challenges [3]. At its core, Smart Agriculture is about making farming more precise, efficient, and responsive to the complex challenges of the 21st century. The concept revolves around the integration of Information and Communication Technologies (ICT) with agricultural practices, allowing for more accurate, timely, and targeted management of farming operations. The fundamental idea is to use data-driven insights to make informed decisions about every aspect of farming, from seed selection and planting to harvesting and postharvest handling. This approach acknowledges the variability within fields and among different crops or livestock, enabling customized management strategies that optimize outcomes while minimizing resource use and environmental impact.

Key components that define Smart Agriculture include data collection through various sensors, satellites, drones, and IoT devices; advanced data analysis using artificial intelligence and machine learning; precision implementation of farm inputs; automation through robotics and autonomous machinery; interconnectivity of smart devices; and sophisticated decision support systems [4]. These elements work together to create an integrated system where various technologies collaborate to provide comprehensive solutions for modern agricultural challenges. Smart Agriculture aims to address several key objectives, including increased productivity, resource efficiency, environmental sustainability, economic viability, food security, and climate resilience. By optimizing inputs and management practices, it seeks to improve crop yields and livestock productivity while reducing waste and improving the efficiency of water, fertilizer, and pesticide use. This approach not only helps to reduce agriculture's environmental footprint but also contributes to global food security by enhancing production capabilities and reducing losses. The benefits of Smart Agriculture extend beyond the farm gate [5]. It has the potential to optimize labor through automation, facilitate knowledge transfer among farmers, and bridge the gap between scientific research and on-farm application. Moreover, by providing detailed data on farm operations and produce quality, it can open new market opportunities through improved traceability and quality assurance.

However, the implementation of Smart Agriculture also faces challenges. These include addressing the digital divide to ensure equitable access to technologies, protecting data privacy and security [6, 7], overcoming the high initial investment costs, developing the necessary skills among farmers and agricultural workers, establishing common standards for data exchange, and addressing ethical considerations related to the impact of automation on rural employment and traditional farming practices. In essence, Smart Agriculture represents a paradigm shift from traditional farming methods that often rely on intuition and general guidelines to a more scientific, data-driven approach. It is not just about individual technologies, but rather about creating an integrated system where various technologies work together to provide a comprehensive solution for modern agricultural challenges [8, 9]. As such, Smart Agriculture is continually evolving, incorporating new technologies and methodologies as they emerge, always with the goal of making farming more efficient, sustainable, and resilient in the face of global challenges like climate change, population growth, and resource scarcity.

12.3 THE HISTORY OF SMART AGRICULTURE

The history of smart agriculture represents a fascinating evolution of farming practices, combining traditional agricultural knowledge with technological advancements. Its roots can be traced back to ancient times, with early innovations like crop rotation and selective breeding laying the groundwork for future developments. The 18th and 19th centuries saw the mechanization of farming and the advent of scientific soil management, setting the stage for more advanced techniques. The true beginnings of precision agriculture emerged in the 1980s and 1990s with the introduction of Global Positioning System (GPS) technology in agriculture, the development of yield monitoring systems for combined harvesters, and early variable rate technology for fertilizer application. This period also saw the emergence of site-specific crop management, including grid soil sampling, variable rate fertilizer application, and the use of Geographic Information Systems (GIS) in agriculture [10]. The early 2000s marked the rise of digital agriculture, characterized by the widespread adoption of GPS-guided tractors and machinery, the development of farm management software, and increased use of satellite imagery for crop monitoring. This era also saw the introduction of the Internet of Things (IoT) in agriculture, with wireless sensor networks for environmental monitoring, smart irrigation systems, and early adoption of RFID technology for livestock tracking. The late 2000s and early 2010s brought big data and cloud computing to agriculture, enabling more sophisticated data analytics and cloud-based farm management platforms. This period also saw the integration of multiple data sources for decision support, paving the way for more advanced applications of technology in farming.

The 2010s marked a significant leap forward with the application of artificial intelligence (AI) and machine learning (ML) in agriculture. This included the development of AI-powered crop and yield prediction models, machine learning for pest and disease detection, and robotic systems for harvesting and weed control [11]. Drone technology also became widespread during this period, used for crop monitoring, mapping, and even spraying and seeding. The late 2010s saw the implementation of blockchain technology in agriculture, primarily for supply chain traceability and smart contracts for agricultural transactions. As we moved into the 2020s, edge computing and 5G networks began enabling faster, more reliable data processing and connectivity in rural areas. Looking to the future, fully autonomous farming systems and swarm robotics for agricultural operations are emerging as the next frontier in smart agriculture. Throughout this evolution, the goal of smart agriculture has consistently been to increase efficiency, reduce environmental impact, and improve crop yields, transforming farming from a largely intuitive practice to a highly data-driven, precision-based industry. It is important to note that the adoption of these technologies has not been uniform across the globe, with developed countries generally at the forefront of smart agriculture adoption while many developing countries are still in the early stages of this technological transformation. The ongoing development of smart agriculture continues to build on this rich history, with new innovations constantly emerging to address the evolving challenges of modern farming [12].

12.4 CHALLENGES IN MODERN AGRICULTURE

Modern agriculture faces a whole series of challenges that undermine its sustainability and effectiveness in supplying food to the world's growing population. As global demand for food increases, farmers face the impact of climate change, which brings irregular weather patterns, extreme temperatures, and unexpected natural disasters. Such climatic fluctuations can wreak havoc on crops, disrupt planting schedules, and lead to lower yields, putting further pressure on food systems. In the following sections, we discuss in detail other challenges that modern agriculture must face.

- *Global Food Security:* The challenge of feeding a growing global population is becoming increasingly urgent. With projections indicating a population of 9.7 billion by 2050, agricultural production must increase dramatically. This isn't just about producing more food, but also ensuring its equitable distribution and reducing waste throughout the supply chain.
- *Climate Change:* Agriculture is both a contributor to and a victim of climate change. Shifting weather patterns are altering growing seasons, while extreme events like droughts, floods, and heatwaves are becoming more frequent and severe. Farmers must adapt to these changing conditions while also working to reduce agriculture's carbon footprint.
- *Resource Scarcity:* Arable land is becoming scarcer due to urbanization, desertification, and soil degradation. Water scarcity is a growing concern in many regions, exacerbated by climate change and overuse. Phosphorus, a critical component in fertilizers, is a finite resource that's being depleted. These challenges necessitate more efficient and sustainable use of available resources.
- *Environmental Impact:* Conventional agriculture has significant environmental costs, including soil erosion, water pollution from pesticides and fertilizers, loss of biodiversity, and deforestation. There's a growing need to develop farming practices that maintain or enhance ecosystem services rather than depleting them.
- *Labor Shortages:* Many countries are experiencing a shortage of agricultural labor due to an aging farming population and rural-to-urban migration.

This trend is particularly pronounced in developed countries, where the average age of farmers is often over 50.

- *Economic Pressures:* Farmers face volatile commodity prices, rising input costs, and often thin profit margins. The need for significant capital investment in new technologies can be a barrier for many, especially smallholder farmers.
- *Pest and Disease Management:* As global trade and climate change introduce new pests and diseases to different regions, managing these threats becomes more complex. Additionally, the overuse of pesticides has led to increased resistance in many pest species.

12.5 THE ROLE OF IoT IN AGRICULTURAL TRANSFORMATION

The integration of Internet of Things (IoT) technology into agriculture marks a pivotal shift in farming practices, heralding a new era of data-driven decision-making and automated processes. This agricultural revolution, often termed "Agriculture 4.0," is reshaping the industry from the ground up, offering solutions to long-standing challenges while opening up new possibilities for sustainability and efficiency [4]. At the heart of this transformation lies a vast network of interconnected devices and sensors that blanket farms in a digital ecosystem. These technologies work in concert to provide farmers with unprecedented insights into their operations. Soil sensors delve deep into the earth, measuring moisture levels, nutrient content, and pH balance with remarkable precision. Overhead, drones equipped with advanced imaging technology sweep across fields, capturing multispectral data that reveals crop health issues invisible to the naked eye. This wealth of data flows into sophisticated analytics platforms, where artificial intelligence and machine learning algorithms sift through the information to uncover patterns and generate actionable insights. Armed with this knowledge, farmers can make highly informed decisions about every aspect of their operations. They can determine the optimal time to plant, irrigate, fertilize, and harvest with a degree of accuracy that was unimaginable just a few decades ago.

Precision agriculture, enabled by IoT, allows for the micromanagement of fields down to the square meter. GPS-guided machinery can plant seeds and apply inputs with centimeter-level accuracy, ensuring that each plant receives exactly what it needs to thrive. This targeted approach not only boosts yields but also significantly reduces waste, leading to more sustainable farming practices and lower environmental impact. In the realm of livestock management, IoT devices are revolutionizing animal husbandry. Smart ear tags monitor the health and behavior of individual animals, alerting farmers to potential issues before they become serious. Automated feeding systems adjust rations based on each animal's nutritional needs and production levels, optimizing feed efficiency and animal welfare. Beyond the farm gate, IoT is transforming the entire agricultural supply chain. Figure 12.1 shows an example carried out to detect the soil moisture content using IoT. Sensors embedded in packaging can track the condition of produce from field to store, ensuring quality and reducing food waste. Blockchain technology, integrated with IoT systems, provides unprecedented traceability, allowing consumers to trace their food back to its source with a simple scan of a QR code. The impact of IoT on agriculture extends to resource

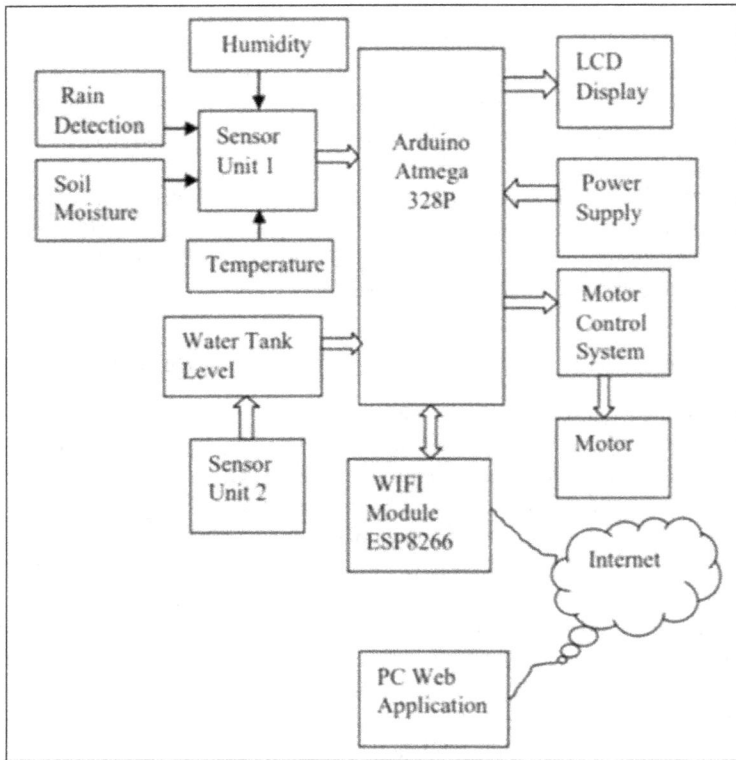

FIGURE 12.1   Example of soil moisture detection using sensors.

management as well. Smart irrigation systems, informed by a network of soil moisture sensors and weather data, can dramatically reduce water usage while improving crop health [13, 14]. Similarly, IoT-enabled energy management systems help farms optimize their power consumption, potentially integrating renewable energy sources more effectively into their operations. As climate change poses increasing challenges to agriculture, the adaptive capabilities offered by IoT become even more crucial. The ability to monitor and respond to changing environmental conditions in real-time helps build resilience into farming systems. Predictive models, fed by a constant stream of IoT data, can help farmers prepare for and mitigate the impacts of extreme weather events or shifting growing seasons. However, the widespread adoption of IoT in agriculture is not without its challenges. The initial investment in technology can be substantial, potentially widening the gap between large, well-funded operations and smaller farms. Issues of data ownership, privacy, and security also loom large, as the vast amount of data generated by IoT systems could be valuable or vulnerable in the wrong hands. Moreover, the successful implementation of IoT in agriculture requires a robust technological infrastructure, including reliable Internet connectivity in rural areas. This presents a significant hurdle in many parts of the world, potentially exacerbating existing inequalities in global agriculture. Despite these challenges, the trajectory of IoT in agriculture points toward a future where farming is more efficient, sustainable, and responsive to global food security needs.

12.6   IoT DEVICES AND SENSOR NETWORKS

The integration of IoT devices and sensor networks in agriculture represents a paradigm shift in farming practices, offering a level of insight and control previously unimaginable. This technological revolution is transforming farms into smart, data-driven operations capable of responding to environmental changes and crop needs with unprecedented precision. At the heart of this transformation lies a diverse array of sensors, each tailored to capture specific aspects of the agricultural ecosystem [15]. Figure 12.2 shows various IoT devices and sensor networks in agriculture. Soil sensors delve beneath the surface, providing real-time data on moisture levels, temperature, pH, and nutrient content at various depths. This granular information allows farmers to understand the exact conditions their crops are experiencing and make informed

FIGURE 12.2   Various IoT devices and sensor networks in agriculture.

decisions about irrigation and fertilization. Complementing these underground sensors are sophisticated weather stations scattered across fields. These compact yet powerful devices measure a range of atmospheric conditions, including temperature, humidity, rainfall, wind speed and direction, and solar radiation. By providing hyperlocal weather data, these stations enable farmers to anticipate microclimatic changes that can significantly impact crop health and yield. Plant sensors take monitoring to the individual crop level. Attached directly to plants, these devices can measure leaf temperature, chlorophyll content, and even minute changes in stem diameter. This data provides invaluable insights into plant health, stress levels, and growth patterns, allowing for early detection of issues like disease or nutrient deficiencies.

Aerial sensors, mounted on drones or satellites, offer a bird's-eye view of crop conditions. Equipped with multispectral and hyperspectral cameras, these sensors can assess crop health across vast areas, detect pest infestations, and estimate yields with remarkable accuracy. This technology enables farmers to identify and address issues in specific areas of their fields, optimizing resource allocation and minimizing crop losses. In livestock farming, wearable sensors are revolutionizing animal husbandry. These devices, often in the form of smart ear tags or collars, can track an animal's location, monitor activity levels, detect changes in body temperature, and even analyze rumination patterns in cattle. This constant stream of data allows farmers to quickly identify health issues, optimize feeding schedules, and improve overall herd management. The true power of these sensors lies in their interconnectedness. Data from individual sensors is aggregated and transmitted through a hierarchical network structure. At the field level, local gateways collect information from multiple sensors. These gateways then relay the data to farm-level systems, which may perform initial processing before sending it to cloud-based platforms for comprehensive analysis [16].

This network relies on a combination of communication technologies to ensure reliable data transmission. Short-range protocols like Bluetooth or ZigBee facilitate communication between nearby devices, while Wi-Fi networks handle higher bandwidth applications within limited areas. For broader coverage, cellular networks (3G/4G/5G) enable real-time data transmission across larger distances. In remote areas without reliable terrestrial network access, satellite communication ensures that even the most isolated farms can benefit from IoT technology. Once collected, this vast amount of data is processed using advanced analytics, often leveraging machine learning and artificial intelligence algorithms. These systems can identify patterns, predict trends, and generate actionable insights that farmers can access through user-friendly dashboards on computers or mobile devices. This allows for data-driven decision-making from anywhere, at any time. The applications of this technology are far-reaching. Precision agriculture, enabled by IoT, allows farmers to apply water, fertilizers, and pesticides with incredible accuracy. Instead of treating entire fields uniformly, farmers can now tailor their approach to the specific needs of different areas or even individual plants. This not only optimizes resource use and reduces waste but also minimizes environmental impact.

Automated systems take this precision one step further. In greenhouses, climate control systems can automatically regulate temperature, humidity, and lighting to create optimal growing conditions [17, 18]. Variable-rate applicators can adjust fertilizer amounts on-the-go based on soil sensor data and GPS location. The benefits of

IoT in agriculture extend beyond the farm gate. These systems can enhance traceability throughout the supply chain, providing valuable data on the journey of food from field to table. This not only improves food safety but also allows consumers to make more informed choices about the products they purchase. As climate change poses increasing challenges to agriculture, the adaptive capabilities offered by IoT become even more crucial. The ability to monitor and respond to changing environmental conditions in real time helps build resilience into farming systems. Predictive models, fed by a constant stream of IoT data, can help farmers prepare for and mitigate the impacts of extreme weather events or shifting growing seasons. However, the widespread adoption of IoT in agriculture is not without challenges. The initial investment in technology can be substantial, potentially widening the gap between large, well-funded operations and smaller farms. Issues of data ownership, privacy, and security also need to be carefully addressed as the vast amount of data generated by these systems is both valuable and potentially vulnerable.

Despite these challenges, the trajectory of IoT in agriculture points toward a future where farming is more efficient, sustainable, and responsive to global food security needs. As these technologies continue to evolve and become more accessible, they have the potential to empower farmers of all scales to produce more food with fewer resources while minimizing environmental impact. This technological revolution in agriculture is not just changing how we farm, it is also reshaping our relationship with food production and the natural world, paving the way for a more sustainable and food-secure future.

12.6.1 SOIL AND CROP SENSORS

Soil and crop sensors are revolutionary tools in modern agriculture, providing farmers with unprecedented insights into their fields and crops. These sophisticated devices form the foundation of precision agriculture, enabling data-driven decision-making that optimizes resource use and enhances crop yields. Soil sensors delve beneath the surface, measuring crucial parameters such as moisture content, temperature, pH levels, electrical conductivity, and nutrient availability [19].

- *Nutrient sensors* are specialized molecular mechanisms within cells that detect and respond to changes in nutrient availability. These sensors play a crucial role in maintaining cellular homeostasis and adapting metabolism to varying environmental conditions. They monitor intracellular and extracellular levels of various nutrients, including glucose, amino acids, and lipids, translating this information into cellular responses by adjusting metabolic pathways and gene expression. Common types of nutrient sensors include glucose sensors (e.g., AMPK, mTOR), amino acid sensors (e.g., GCN2, mTORC1), and lipid sensors (e.g., PPARs, SREBP). These sensors typically function through direct binding of nutrients, indirect detection of metabolic by-products, or monitoring energy status, such as the AMP:ATP ratio. Upon detecting nutrient changes, they activate or inhibit various signaling cascades, leading to altered gene expression, modulation of metabolic enzyme activity, and changes in nutrient uptake and utilization. Nutrient sensors help cells adapt to fluctuating nutrient availability, regulate growth and proliferation, manage energy expenditure, and coordinate with whole-body metabolism.

- *Plant health monitors* are advanced technological tools designed to assess and track the overall well-being of plants. By continuously collecting and analyzing this data, plant health monitors provide real-time insights into a plant's condition, allowing growers to identify potential issues early on. This information enables timely interventions, such as adjusting watering schedules or addressing nutrient deficiencies, ultimately promoting optimal plant growth and crop yields. Plant health monitors are becoming increasingly important in modern agriculture and gardening, helping to optimize resource use and improve plant management practices.

12.6.2 Weather Monitoring Systems

Weather Monitoring Systems are sophisticated technological setups designed to collect, analyze, and report various atmospheric conditions. These systems typically consist of multiple sensors and instruments that measure parameters such as temperature, humidity, air pressure, wind speed and direction, precipitation, and solar radiation. The data gathered is processed and often transmitted in real time to central databases or weather stations.

- *Automated weather stations* are self-contained systems designed to collect and transmit meteorological data without human intervention [20, 21]. These stations typically include a suite of sensors measuring various weather parameters such as temperature, humidity, wind speed and direction, precipitation, and atmospheric pressure. The data is recorded at regular intervals and can be transmitted wirelessly to a central database or accessed remotely. Automated weather stations are valuable for providing localized, real-time weather information, which is crucial for agricultural decision-making, urban planning, and environmental monitoring.
- *Micro climate sensors* are specialized devices that measure environmental conditions in small, specific areas. These sensors are designed to capture data on a much finer scale than traditional weather stations, focusing on the unique conditions within plant canopies, near soil surfaces, or in other localized environments. They can measure factors like leaf wetness, soil temperature, and light intensity at various heights within a crop. This detailed information helps farmers and researchers understand the subtle environmental variations that can significantly impact plant growth and crop performance.

12.6.3 Livestock Tracking Devices

Livestock tracking devices are technologies used to monitor and manage animal populations in agricultural settings. These devices help farmers keep track of their animals' locations, movements, and sometimes health status. The two main types of livestock tracking devices are RFID tags and GPS collars.

- *RFID (Radio-Frequency Identification) tags for animal identification* are small electronic devices attached to animals, typically as ear tags or implants. These tags contain unique identification codes that can be read by RFID scanners [22]. When an animal passes near a scanner, its identity is recorded, allowing for automated tracking of individual animals within a herd. RFID technology enables efficient record-keeping for health treatments, breeding programs, and regulatory compliance [23].
- *GPS Collars for grazing management* are more advanced tracking devices fitted around an animal's neck. These collars use Global Positioning System technology to record the animal's location at regular intervals. The data can be used to monitor grazing patterns, identify preferred foraging areas, and ensure animals don't stray from designated pastures. GPS collars help in optimizing pasture use, detecting unusual behavior that might indicate health issues, and improving overall herd management [24].

12.6.4 IRRIGATION CONTROL SYSTEMS

Irrigation control systems are technologies designed to manage water distribution in agricultural settings efficiently. These systems aim to optimize water use by delivering the right amount of water to crops at the right time, based on factors like soil moisture, weather conditions, and crop water requirements.

- *Smart sprinklers* are automated irrigation devices that use data from various sources to make intelligent watering decisions. These systems can integrate information from weather forecasts, soil moisture sensors, and preprogrammed schedules to determine when and how much to water. Smart sprinklers can adjust their operation based on real-time conditions, such as skipping a watering cycle if rain is expected or increasing water output during hot, dry periods. This adaptive approach helps conserve water while maintaining optimal growing conditions for plants.
- *Drip irrigation controllers* manage low-flow watering systems that deliver water directly to plant roots. These controllers regulate the timing and duration of water release through a network of tubes and emitters. Advanced drip irrigation controllers can be programmed to deliver precise amounts of water based on crop type, growth stage, and environmental conditions. Many systems allow for remote monitoring and control via smartphone apps or web interfaces, enabling farmers to manage irrigation from anywhere. By delivering water precisely where it is needed, drip irrigation controllers significantly reduce water waste and can improve crop yields by maintaining optimal soil moisture levels [25].

12.7 DATA COLLECTION AND MANAGEMENT

Data collection and management in agriculture involve the systematic gathering, storage, processing, and analysis of information from various farm operations and environmental sources. This process is crucial for making informed decisions, optimizing resource use, and improving overall farm productivity.

12.7.1 Cloud-Based Platforms

Cloud-based platforms in agriculture are online services that provide centralized storage and processing capabilities for farm data. These platforms allow farmers to access their information from any device with an Internet connection, facilitating remote management and collaboration.

- *Data storage and processing* offers farmers a secure and scalable solution for handling large volumes of agricultural data. These systems can store diverse types of information, including sensor readings, weather data, crop yields, and equipment performance metrics. The cloud infrastructure enables powerful data processing capabilities, allowing for complex analyses, pattern recognition, and predictive modeling. This centralized approach helps farmers gain insights from their data without needing to invest in expensive on-site computing hardware [26].
- *Real-time monitoring dashboards* are user-friendly interfaces that display up-to-date information about various aspects of farm operations. These dashboards typically present data through graphs, charts, and maps, making it easy for farmers to quickly assess the current state of their farms. They can show information such as soil moisture levels, weather forecasts, equipment locations, and crop health indicators. By providing a comprehensive overview of farm conditions, these dashboards enable rapid decision-making and proactive management of potential issues [16].

12.7.2 Edge Computing in Agriculture

Edge computing in agriculture refers to the practice of processing data closer to its source, typically on or near the farm, rather than sending all data to a centralized cloud system [27]. This approach can offer several benefits in agricultural applications.

- *On-farm data processing* involves using local computing devices to analyze data from sensors and equipment before sending the results to a central system. This approach can include tasks such as filtering out irrelevant data, performing initial calculations, or triggering immediate action based on predefined rules. By processing data locally, farmers can reduce the amount of information that needs to be transmitted and stored in the cloud, potentially lowering costs and improving system efficiency.
- *Reducing latency and bandwidth usage:* Edge computing can significantly reduce latency – the time delay between data collection and action by processing critical information locally. This is particularly important for time-sensitive operations, such as automated irrigation systems or livestock monitoring. Additionally, by only sending processed or summarized data to the cloud, edge computing can substantially reduce bandwidth usage [28]. This is especially beneficial in rural areas where Internet connectivity may be limited or expensive. The reduced data transmission also helps in scenarios where real-time decision-making is crucial, as it allows for faster responses to changing conditions on the farm [29].

12.8   DATA ANALYTICS AND FUSION

Data analytics and fusion in agriculture represent a cutting-edge approach to farm management, combining diverse data sources and advanced analytical techniques to provide comprehensive insights [30]. This field is rapidly evolving, leveraging technological advancements to transform traditional farming practices into data-driven, precision agriculture.

12.8.1   GEOSTATISTICAL ANALYSIS

Geostatistical analysis in agriculture is crucial for understanding spatial patterns and relationships in farming landscapes. It helps in creating detailed maps of various agricultural parameters, enabling precise and localized management strategies.

- *Spatial variability modeling* is essential for understanding how different factors vary across a field or region. This technique involves creating detailed maps of soil properties (e.g., pH, nutrient levels, and organic matter content), modeling crop yield variations within fields, and analyzing patterns of pest and disease spread.
- *Kriging and Co-kriging techniques* are advanced interpolation methods used to estimate values at unsampled locations.
  - *Kriging* might be used to create a detailed soil moisture map based on readings from a limited number of soil sensors across a field.
  - *Co-kriging* could enhance this by incorporating related data, such as topography or soil type, to improve the accuracy of the moisture estimates.

12.8.2   TIME SERIES MODELING

Time series modeling in agriculture focuses on analyzing temporal patterns to make predictions and inform decision-making.

- *Crop growth prediction* typically incorporate daily weather data (temperature, precipitation, and solar radiation), soil conditions (moisture, nutrients), and crop-specific growth parameters. These models can simulate crop development stages, predict biomass accumulation, and estimate water and nutrient needs throughout the growing season. For instance, a corn growth model might predict when the crop will reach various growth stages, helping farmers time their management activities more precisely.
- *Yield forecasting* models have become increasingly sophisticated, incorporating in historical yield data, current season weather patterns, remote sensing data (e.g., NDVI from satellite imagery), and crop model outputs. Advanced yield forecasting systems might use ensemble methods, combining multiple models to improve prediction accuracy. These forecasts are valuable not just for individual farmers but also for agricultural commodity markets and food security planning at regional or national levels.

12.8.3 MULTISOURCE DATA INTEGRATION

Multisource data integration is at the heart of modern precision agriculture, bringing together diverse data streams to create a holistic view of agricultural systems.

- *Sensor fusion algorithms* might involve combining data from soil moisture probes, weather stations, and crop canopy sensors to optimize irrigation scheduling; integrating readings from multispectral cameras, thermal sensors, and LiDAR to assess crop health and structure; merging data from GPS-enabled farm equipment with yield monitors to create high-resolution productivity maps. These fusion algorithms often need to handle data with different temporal and spatial resolutions, requiring sophisticated statistical and computational techniques.
- *Combining satellite and ground-based data:* The integration of satellite and ground-based data is transforming large-scale agricultural monitoring such that satellite imagery provides broad coverage and regular updates on crop conditions, often using vegetation indices like NDVI or EVI. Ground-based data from weather stations, soil sensors, and field observations provide detailed, point-based measurements. Combining these sources allows for calibration and validation of satellite-derived products, improving their accuracy and relevance for local conditions

12.9 ARTIFICIAL INTELLIGENCE IN AGRICULTURE

Artificial intelligence (AI) is revolutionizing agriculture by providing sophisticated tools for data analysis, prediction, and decision-making. This section explores how AI, particularly machine learning, computer vision, and reinforcement learning, is being applied to solve complex agricultural challenges and transform farming practices [31].

12.9.1 MACHINE LEARNING FOR CROP PREDICTION

Machine learning (ML) algorithms are analyzing vast amounts of agricultural data to make predictions about crop yield and quality, offering farmers unprecedented insights into their operations.

- *Yield estimation models* use ML algorithms to predict crop yields based on factors such as historical yield data, weather patterns, soil conditions, crop management practices, satellite imagery, and genetic information. These models provide early yield forecasts, identify influential factors, assess various scenarios, and aid in risk management.
- *Crop quality prediction:* ML predicts crop quality by considering growing conditions, harvest timing, postharvest handling, genetic factors, and pest and disease pressure. Techniques like convolutional neural networks and Random Forests analyze data to predict visual quality, nutritional content, flavor profiles, storage potential, and processing characteristics [32].

12.9.2 Computer Vision for Plant Health

Computer Vision is revolutionizing how farmers monitor plant health and detect issues early, transforming crop protection and nutrition management.

- *Disease detection:* Advanced computer vision systems analyze leaf images to identify disease symptoms, monitor crop canopy for stress signs, track disease progression, and differentiate between multiple stressors. These systems offer early detection, large-scale monitoring, consistent results, rare disease identification, and quantitative assessment.
- *Weed identification:* Computer vision systems for weed control identify weed species, map distributions, guide precision spraying, monitor herbicide resistance, and support mechanical weeding. Benefits include reduced herbicide use, improved crop yields, environmental protection, and labor savings [33].

12.9.3 Reinforcement Learning for Optimization

Reinforcement learning (RL) is optimizing various agricultural processes, offering adaptive and innovative solutions to complex farming challenges.

- *Irrigation scheduling:* RL systems for irrigation learn from historical data and current conditions, balancing water conservation with crop needs. They can control individual sprinklers, integrate multiple data sources, optimize for various objectives, adapt to changing conditions, and anticipate future needs.
- *Pest control strategies:* RL is developing adaptive pest control strategies, learning optimal timing for pesticide application, developing integrated pest management strategies, and adapting to changing pest populations. Benefits include reduced pesticide use, resistance management, ecosystem preservation, and long-term sustainability. As these AI technologies advance and integrate with other aspects of farm management, they are shaping the future of agriculture. By providing farmers with powerful tools to increase productivity, sustainability, and resilience, AI is helping meet the growing global demand for food while minimizing environmental impact. The ongoing development of these technologies promises even more sophisticated and effective agricultural solutions in the coming years.

12.10 PRECISION AGRICULTURE APPLICATIONS

Various IoT solutions used in agricultural problems are shown in Figure 12.3. Precision agriculture represents a paradigm shift in farming, leveraging advanced technologies to optimize crop management and resource utilization. This approach aims to enhance efficiency, sustainability, and productivity in agricultural practices [34].

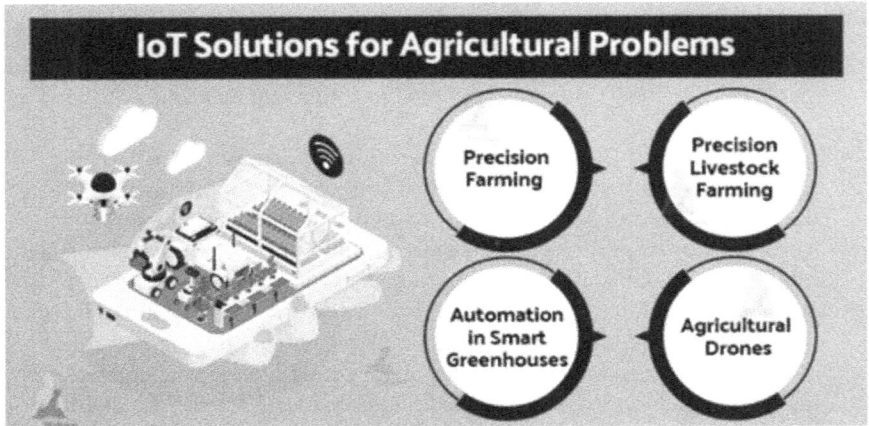

FIGURE 12.3   Solutions using IoT and AI in farming.

12.10.1 VARIABLE RATE TECHNOLOGY

VRT is a cornerstone of precision agriculture, allowing farmers to apply inputs at varying rates across a field, tailoring management to specific needs of different areas.

- *Fertilizer application:* VRT in fertilizer application uses detailed soil nutrient maps and real-time crop health data to apply fertilizers precisely where and when they are needed. This approach significantly reduces waste and environmental impact while optimizing crop nutrition.
  - *Example of this application:* A farmer uses grid soil sampling to collect data on nitrogen levels across a field. This data is combined with satellite imagery showing crop vigor (NDVI). A machine learning model, trained on historical data, predicts nitrogen needs for different areas. The field is then segmented into management zones using k-means clustering. Finally, a prescription map is generated using kriging interpolation, guiding the variable-rate fertilizer applicator.
- *Pesticide spraying:* VRT for pesticide application uses pest pressure maps, weed distribution data, and crop health information to target spraying. This approach can significantly reduce pesticide use while maintaining effective pest control.
  - *Example of this application:* Drones equipped with high-resolution cameras survey a field. Images are processed using a CNN to identify and map weed patches. This data, combined with weather forecasts and historical pest pressure data, is fed into a Random Forest model to predict high-risk areas for pest outbreaks. A variable-rate sprayer uses this information to apply pesticides only where needed, adjusting in real time using PID control based on ground speed and wind conditions [35].

12.10.2 SITE-SPECIFIC CROP MANAGEMENT

This approach moves beyond the one-size-fits-all model of farming, tailoring management practices to specific areas within a field based on their unique characteristics.

- *Zonal management:* Fields are divided into management zones based on factors like soil type, topography, yield history, and other relevant parameters. Each zone is managed as a distinct unit, optimizing inputs and practices for its specific conditions.
  - *Example of this application:* A farmer collects multiple layers of data: soil EC (electrical conductivity) maps, historical yield data, and topography. PCA is used to identify the most important factors. Fuzzy C-means clustering is then applied to delineate management zones. Moran's I is calculated to ensure zones are spatially coherent. A Bayesian network-based decision support system then provides management recommendations for each zone, considering crop type, weather forecasts, and economic factors.
- *Prescription maps* guide variable rate applications by providing a detailed, georeferenced plan for input application across a field. They integrate multiple data sources to optimize resource use.
  - *Example of this application:* Soil nutrient data from lab analysis is combined with EC maps and crop health indices from satellite imagery. Co-kriging is used to create continuous maps of each variable. These maps are then fused using Dempster–Shafer theory to account for the reliability of each data source. Finally, a genetic algorithm optimizes the fertilizer prescription, balancing crop needs, input costs, and environmental constraints [36].

12.10.3 SMART IRRIGATION SYSTEMS

Smart irrigation systems optimize water use by considering multiple factors, including soil moisture, weather conditions, crop water needs, and even economic considerations.

- *Soil moisture-based irrigation:* These systems use a network of soil moisture sensors to determine when and how much to irrigate, ensuring crops receive optimal water without waste [37, 38].
  - *Example of this application:* A field is equipped with a network of soil moisture sensors, placed using spatial simulated annealing to optimize coverage. Raw sensor data is filtered using a Kalman filter to reduce noise. A water balance model, updated with real-time sensor data, predicts soil moisture depletion. When predicted soil moisture drops below a threshold, irrigation is triggered. An artificial neural network, trained on historical data, fine-tunes the irrigation amount based on crop stage, soil type, and recent weather patterns.

- *Weather-responsive watering:* These systems integrate weather data and forecasts to adjust irrigation schedules, anticipating rainfall or high evapotranspiration conditions.
  - *Example of this application:* A smart irrigation system receives local weather data and forecasts. It uses an artificial neural network to estimate daily evapotranspiration. This is fed into a crop growth model (e.g., DSSAT) to predict crop water needs. A Model Predictive Control Algorithm then optimizes the irrigation schedule for the next several days, considering predicted rainfall, crop water needs, energy costs (for pumping), and any water use restrictions. The system continuously updates its plan as new weather data becomes available [39].

12.11   AGRICULTURAL ROBOTICS AND AUTOMATION

Agricultural robotics and automation are transforming traditional farming practices, offering solutions to labor shortages, increasing efficiency, and enabling more precise management of crops and resources.

12.11.1   AUTONOMOUS TRACTORS AND MACHINERY

Autonomous tractors and machinery represent a significant advancement in farm automation, capable of performing various tasks with minimal human intervention.

- *Self-driving technologies* in agriculture utilize a combination of GPS, sensors, and artificial intelligence to navigate fields and perform tasks autonomously.
  - *Example of this application:* An autonomous tractor uses RTK GPS for precise positioning. Its path through the field is planned using the A* algorithm, optimizing for efficiency and avoiding known obstacles. LiDAR and cameras continuously scan for unexpected obstacles, with a CNN-based object detection system classifying potential hazards. The tractor uses SLAM to build and update a detailed map of the field, including information on soil compaction and crop rows. A Reinforcement Learning Algorithm allows the tractor to adapt its speed and wheel slip based on changing soil conditions, optimizing traction and minimizing soil damage.
- *Robotic implements:* Smart implements can adjust their operations based on real-time field conditions, allowing for more precise and efficient farm operations.
  - *Example of this application:* A robotic planter uses spectral sensors to analyze soil conditions in real time. This data feeds into a Model Predictive Control system that continuously adjusts planting depth and spacing. A genetic algorithm optimizes seed placement based on soil fertility maps, expected rainfall, and crop-specific requirements. For a combined planter–sprayer unit, a deep learning model identifies weeds in real time, allowing for precise, targeted herbicide application only where needed.

12.11.2 ROBOTIC HARVESTING SYSTEMS

Robotic harvesting systems are addressing labor shortages and enabling more efficient and gentle harvesting of crops [40].

- *Fruit picking robots:* These robots use advanced vision systems and sophisticated arm control to identify and gently pick ripe fruit.
  - *Example of this application:* A fruit picking robot scans an apple tree using multiple cameras [41]. A CNN detects and localizes ripe apples, while a Structure from Motion Algorithm creates a 3D model of the tree. An RRT algorithm plans the optimal path for the robotic arm to reach each apple. The gripper, made of soft, compliant materials, gently grasps each apple using force feedback control. A spectrometer on the gripper analyzes the apple's surface, with a Random Forest classifier determining optimal ripeness for picking.
- *Grain harvesting automation:* Automated grain harvesting systems optimize the harvesting process, reducing waste and improving grain quality.
  - *Example of this application:* An automated combine harvester uses computer vision for precise row following. Its threshing system uses a fuzzy logic controller to continuously adjust rotor speed and concave clearance based on crop conditions and feedback from loss sensors. A NIRS system analyzes grain quality in real time, with results feeding into a Partial Least Squares Regression model to estimate protein content. GPS-linked yield sensors create a real-time yield map, interpolated using Kriging. Meanwhile, vibration sensors and temperature monitors feed data into a Random Forest model that predicts potential equipment failures, allowing for preventive maintenance.

12.11.3 DRONE-BASED CROP MONITORING

Drones are revolutionizing crop monitoring, providing high-resolution, timely data on crop health and field conditions [42].

- *Multispectral imaging:* Drones equipped with multispectral cameras capture data across various light spectrums, providing insights into crop health, stress, and nutrient status.
  - *Example of this application:* A drone equipped with a 5-band multispectral camera flies a preprogrammed path over a field. Images are stitched together using a Structure from Motion Algorithm to create an orthomosaic. Radiometric calibration is performed using ground targets and the Empirical Line Method. Multiple vegetation indices (NDVI, NDRE) are calculated. A Random Forest classifier, trained on historical data, uses these indices to identify areas of crop stress. The resulting stress map is used to guide scouting efforts and inform variable rate applications of fertilizer or pesticides.

- *Crop spraying drones:* Drones capable of carrying and applying pesticides or fertilizers offer a precise, flexible alternative to traditional spraying methods.
  - *Example of this application:* A swarm of spraying drones receives a prescription map for pesticide application. Each drone uses RTK GPS for precise navigation, with a SLAM system for obstacle avoidance. Spray nozzles, optimized using computational fluid dynamics, provide even coverage. Flow rates are continuously adjusted based on flight speed and the prescription map [43]. The drones coordinate their movements using a distributed algorithm to ensure complete coverage without overlap. After spraying, the drones transmit detailed application maps, providing a record of exactly where, when, and how much pesticide was applied.

12.12   FUTURE TRENDS AND OPPORTUNITIES

As technology continues to evolve, new trends and opportunities are emerging in the field of smart agriculture, promising to further revolutionize farming practices.

- *5G and Advanced Connectivity:* The rollout of 5G networks and other advanced connectivity solutions is set to dramatically enhance the capabilities of smart farming systems.
- *Improved Rural Internet Access:* Expanding high-speed Internet access to rural areas will enable more farmers to adopt and fully utilize smart farming technologies.
- *Low-Latency Applications:* 5G's low latency enables new applications requiring real-time data processing and control.
- *Blockchain for Supply Chain Traceability:* Blockchain technology is poised to transform agricultural supply chains, enhancing transparency, traceability, and trust.
- *Food Safety and Quality Assurance:* Blockchain can create immutable records of food production, processing, and distribution, enhancing food safety and quality assurance.
- *Transparent Farm-to-Table Tracking:* Blockchain enables consumers to trace their food's journey from farm to table, promoting transparency and informed choices.
- *Integration with Smart City Systems:* As urban areas evolve into smart cities, there are increasing opportunities for integration with smart agriculture systems.
- *Urban Farming Technologies:* Smart technologies are enabling efficient food production in urban environments.
- *Sustainable Food Systems:* Integration of smart farming with urban systems can create more sustainable and efficient food production and distribution networks.

12.13 CONCLUSION AND OUTLOOK

The rapid advancement of IoT and AI technologies in agriculture is paving the way for a new era of farming that is more efficient, sustainable, and responsive to global challenges. In conclusion, the future of agriculture is inextricably linked with the advancement of IoT and AI technologies. As these technologies continue to evolve and become more integrated with farming practices, they promise to address many of the pressing challenges facing global agriculture. However, realizing this potential will require not only technological innovation but also careful consideration of environmental, social, and ethical implications. The path forward will likely involve a balance between high-tech solutions and traditional agricultural wisdom, aiming to create a food system that is not only productive and efficient but also sustainable, resilient, and equitable.

REFERENCES

[1] M. R. M. Kassim, "IoT Applications in Smart Agriculture: Issues and Challenges," in *2020 IEEE Conference on Open Systems (ICOS)*, Nov. 2020, pp. 19–24. https://doi.org/10.1109/ICOS50156.2020.9293672.

[2] E. Kannan, C. M. B. M J, A. D. S, R. N. N, A. Begum, and H. D, "Deep Learning Techniques Advancements in Apple Leaf Disease Detection," *Procedia Computer Science*, vol. 235, pp. 713–722, 2024. https://doi.org/10.1016/j.procs.2024.04.068.

[3] K. El-Moustaqim, J. Mabrouki, M. Azrour, M. Hadine, and D. Hmouni, "Enabling Smart Agriculture through Integrating the Internet of Things in Microalgae Farming for Sustainability," in *Smart Internet of Things for Environment and Healthcare*, M. Azrour, J. Mabrouki, A. Alabdulatif, A. Guezzaz, and F. Amounas, Eds., Cham: Springer Nature Switzerland, 2024, pp. 209–222. https://doi.org/10.1007/978-3-031-70102-3_15.

[4] R. Dagar, S. Som, and S. K. Khatri, "Smart Farming – IoT in Agriculture," in *2018 International Conference on Inventive Research in Computing Applications (ICIRCA)*, 2018, pp. 1052–1056. https://doi.org/10.1109/ICIRCA.2018.8597264.

[5] A. Chandra, K. E. McNamara, and P. Dargusch, "Climate-Smart Agriculture: Perspectives and Framings," *Climate Policy*, vol. 18, no. 4, pp. 526–541, 2018. https://doi.org/10.1080/14693062.2017.1316968.

[6] S. Dargaoui, et al., "IoT-Driven Smart Agriculture: Security Issues and Authentication Schemes Classification," in *Proceeding of the International Conference on Connected Objects and Artificial Intelligence (COCIA2024)*, Y. Mejdoub and A. Elamri, Eds., Cham: Springer Nature Switzerland, 2024, pp. 61–66. https://doi.org/10.1007/978-3-031-70411-6_10.

[7] S. Dargaoui, et al., "Internet-of-Things-Enabled Smart Agriculture: Security Enhancement Approaches," in *2024 4th International Conference on Innovative Research in Applied Science, Engineering and Technology (IRASET)*, May 2024, pp. 1–5. https://doi.org/10.1109/IRASET60544.2024.10548705.

[8] M. Mohy-eddine, A. Guezzaz, S. Benkirane, and M. Azrour, "IoT-Enabled Smart Agriculture: Security Issues and Applications," in *Artificial Intelligence and Smart Environment: ICAISE'2022*, Springer, 2023, pp. 566–571.

[9] J. Mabrouki, et al., "Smart System for Monitoring and Controlling of Agricultural Production by the IoT," in *IoT and Smart Devices for Sustainable Environment*, Springer, 2022, pp. 103–115.

[10] D. Li, "An Overview of Earth Observation and Geospatial Information Service," in *Geospatial Technology for Earth Observation*, D. Li, J. Shan, and J. Gong, Eds., Boston, MA: Springer, 2009, pp. 1–25. https://doi.org/10.1007/978-1-4419-0050-0_1.

[11] T. A. Khoa, M. M. Man, T.-Y. Nguyen, V. Nguyen, and N. H. Nam, "Smart Agriculture Using IoT Multi-Sensors: A Novel Watering Management System," *Journal of Sensor and Actuator Networks*, vol. 8, no. 3, Art. no. 3, 2019. https://doi.org/10.3390/jsan8030045.

[12] K. A. Patil and N. R. Kale, "A Model for Smart Agriculture Using IoT," in *2016 International Conference on Global Trends in Signal Processing, Information Computing and Communication (ICGTSPICC)*, 2016, pp. 543–545. https://doi.org/10.1109/ICGTSPICC.2016.7955360.

[13] M. Azrour, et al., "A Survey of Machine and Deep Learning Applications in the Assessment of Water Quality," in *World Sustainability Series*, vol. Part F2854, 2024, pp. 471–483. https://doi.org/10.1007/978-3-031-56292-1_38.

[14] J. Mabrouki, M. Azrour, and S. El Hajjaji, "Use of Internet of Things for Monitoring and Evaluation Water's Quality: Comparative Study," *International Journal of Cloud Computing*, vol. 10, no. 5–6, pp. 633–644, 2021.

[15] J. Muangprathub, N. Boonnam, S. Kajornkasirat, N. Lekbangpong, A. Wanichsombat, and P. Nillaor, "IoT and Agriculture Data Analysis for Smart Farm," *Computers and Electronics in Agriculture*, vol. 156, pp. 467–474, 2019. https://doi.org/10.1016/j.compag.2018.12.011.

[16] S. Namani and B. Gonen, "Smart Agriculture Based on IoT and Cloud Computing," in *2020 3rd International Conference on Information and Computer Technologies (ICICT)*, 2020, pp. 553–556. https://doi.org/10.1109/ICICT50521.2020.00094.

[17] N. Ben-Lhachemi, M. Benchrifa, S. Nasrdine, J. Mabrouki, M. Slaoui, and M. ade Azrour, "Effect of IoT Integration in Agricultural Greenhouses," in *Technical and Technological Solutions Towards a Sustainable Society and Circular Economy*, J. Mabrouki and A. Mourade, Eds., Cham: Springer Nature Switzerland, 2024, pp. 435–445. https://doi.org/10.1007/978-3-031-56292-1_35.

[18] M. Benzyane, M. Azrour, I. Zeroual, and S. Agoujil, "State-of-the-Art Methods for Dynamic Texture Classification: A Comprehensive Review," in *Sustainable and Green Technologies for Water and Environmental Management*, M. Azrour, J. Mabrouki, and A. Guezzaz, Eds., Cham: Springer Nature Switzerland, 2024, pp. 1–13. https://doi.org/10.1007/978-3-031-52419-6_1.

[19] S. Getahun, H. Kefale, and Y. Gelaye, "Application of Precision Agriculture Technologies for Sustainable Crop Production and Environmental Sustainability: A Systematic Review," *Scientific World Journal*, vol. 2024, no. 1, p. 2126734, 2024. https://doi.org/10.1155/2024/2126734.

[20] J. Mabrouki, M. Azrour, D. Dhiba, Y. Farhaoui, and S. E. Hajjaji, "IoT-Based Data Logger for Weather Monitoring Using Arduino-Based Wireless Sensor Networks with Remote Graphical Application and Alerts," *Big Data Mining and Analytics*, vol. 4, no. 1, pp. 25–32, 2021. https://doi.org/10.26599/BDMA.2020.9020018.

[21] K. S. Ram and A. N. P. S. Gupta, "IoT Based Data Logger System for Weather Monitoring Using Wireless Sensor Networks," *International Journal of Engineering Trends and Technology*, vol. 32, no. 2, pp. 71–75, 2016. https://doi.org/10.14445/22315381/IJETT-V32P213.

[22] M. Nazar Dawood, et al., "Design of Traceability Platform for Animal Husbandry Products Supply Chain Based on RFID Internet of Things," in *2024 International Conference on Smart Systems for Electrical, Electronics, Communication and Computer Engineering (ICSSEECC)*, 2024, pp. 701–706. https://doi.org/10.1109/ICSSEECC61126.2024.10649535.

[23] Y. Chu, E. Kepros, B. Avireni, S. K. Ghosh, and P. Chahal, "RF Energy Harvesting Hybrid RFID Based Sensors for Smart Agriculture Applications," in *2024 IEEE 74th Electronic Components and Technology Conference (ECTC)*, 2024, pp. 2267–2271. https://doi.org/10.1109/ECTC51529.2024.00385.

[24] A. Mitra, et al., "Smart Agriculture: A Comprehensive Overview," *SN Computer Science*, vol. 5, no. 8, p. 969, 2024. https://doi.org/10.1007/s42979-024-03319-w.

[25] A. E G and G. J. Bala, "IoT and ML-Based Automatic Irrigation System for Smart Agriculture System," *Agronomy Journal*, vol. 116, no. 3, pp. 1187–1203, 2024. https://doi.org/10.1002/agj2.21344.

[26] A. Morchid, I. G. Muhammad Alblushi, H. M. Khalid, R. El Alami, S. R. Sitaramanan, and S. M. Muyeen, "High-Technology Agriculture System to Enhance Food Security: A Concept of Smart Irrigation System Using Internet of Things and Cloud Computing," *Journal of the Saudi Society of Agricultural Sciences*, 2024. https://doi.org/10.1016/j.jssas.2024.02.001.

[27] M. J. O'Grady, D. Langton, and G. M. P. O'Hare, "Edge Computing: A Tractable Model for Smart Agriculture?" *Artificial Intelligence in Agriculture*, vol. 3, pp. 42–51, 2019. https://doi.org/10.1016/j.aiia.2019.12.001.

[28] X. Li, L. Zhu, X. Chu, and H. Fu, "Edge Computing-Enabled Wireless Sensor Networks for Multiple Data Collection Tasks in Smart Agriculture," *Journal of Sensors*, vol. 2020, no. 1, p. 4398061, 2020. https://doi.org/10.1155/2020/4398061.

[29] A. Elsayed and M. Abouhawwash, "An Effective Model for Selecting the Best Cloud Platform for Smart Farming in Smart Cities: A Case Study," *Optimization in Agriculture*, vol. 1, pp. 66–80, 2024. https://doi.org/10.61356/j.oia.2024.1202.

[30] X. Pham and M. Stack, "How Data Analytics Is Transforming Agriculture," *Business Horizons*, vol. 61, no. 1, pp. 125–133, 2018. https://doi.org/10.1016/j.bushor.2017.09.011.

[31] A. Nigam, S. Garg, A. Agrawal, and P. Agrawal, "Crop Yield Prediction Using Machine Learning Algorithms," in *2019 Fifth International Conference on Image Information Processing (ICIIP)*, 2019, pp. 125–130. https://doi.org/10.1109/ICIIP47207.2019.8985951.

[32] M. Kalimuthu, P. Vaishnavi, and M. Kishore, "Crop Prediction Using Machine Learning," in *2020 Third International Conference on Smart Systems and Inventive Technology (ICSSIT)*, 2020, pp. 926–932. https://doi.org/10.1109/ICSSIT48917.2020.9214190.

[33] S. S. Chouhan, U. P. Singh, and S. Jain, "Applications of Computer Vision in Plant Pathology: A Survey," *Archives of Computational Methods in Engineering*, vol. 27, no. 2, pp. 611–632, 2020. https://doi.org/10.1007/s11831-019-09324-0.

[34] M. Dholu and K. A. Ghodinde, "Internet of Things (IoT) for Precision Agriculture Application," in *2018 2nd International Conference on Trends in Electronics and Informatics (ICOEI)*, 2018, pp. 339–342. https://doi.org/10.1109/ICOEI.2018.8553720.

[35] N. Delavarpour, C. Koparan, J. Nowatzki, S. Bajwa, and X. Sun, "A Technical Study on UAV Characteristics for Precision Agriculture Applications and Associated Practical Challenges," *Remote Sensing*, vol. 13, no. 6, Art. no. 6, 2021. https://doi.org/10.3390/rs13061204.

[36] I. Attri, L. K. Awasthi, and T. P. Sharma, "Machine Learning in Agriculture: A Review of Crop Management Applications," *Multimedia Tools and Applications*, vol. 83, no. 5, pp. 12875–12915, 2024. https://doi.org/10.1007/s11042-023-16105-2.

[37] K. Arunadevi, M. Singh, M. Khanna, A. K. Mishra, and V. K. Prajapati, "Moisture Sensor-Based Irrigation Scheduling to Improve Water Productivity in Agriculture," in *Recent Advancements in Sustainable Agricultural Practices: Harnessing Technology for Water Resources, Irrigation and Environmental Management*, Yasheshwar, A. K. Mishra, and M. Kumar, Eds., Singapore: Springer Nature, 2024, pp. 113–131. https://doi.org/10.1007/978-981-97-2155-9_6.

[38] B. Gebeyhu, S. Dagalo, and M. Muluneh, "Soil Moisture-Based Irrigation Interval and Irrigation Performance Evaluation: In the Case of Lower Kulfo Catchment, Ethiopia," *Heliyon*, vol. 10, no. 16, 2024. https://doi.org/10.1016/j.heliyon.2024.e36089.

[39] B. Kouhzad, M. R. Yazdani, and M. T. Dastorani, "Evaluating Negarim Microcatchment Efficiency to Conserve Soil Moisture Based on Soil Depth," *Research Square*, 2024. https://doi.org/10.21203/rs.3.rs-4704859/v1.

[40] T. Jin and X. Han, "Robotic Arms in Precision Agriculture: A Comprehensive Review of the Technologies, Applications, Challenges, and Future Prospects," *Computers and Electronics in Agriculture*, vol. 221, p. 108938, 2024. https://doi.org/10.1016/j.compag.2024.108938.

[41] K. Zhang, K. Lammers, P. Chu, Z. Li, and R. Lu, "An Automated Apple Harvesting Robot – From System Design to Field Evaluation," *Journal of Field Robotics*, vol. 41, no. 7, pp. 2384–2400, 2024. https://doi.org/10.1002/rob.22268.

[42] P. Kar and S. Chowdhury, "IoT and Drone-Based Field Monitoring and Surveillance System," in *Artificial Intelligence Techniques in Smart Agriculture*, S. S. Chouhan, A. Saxena, U. P. Singh, and S. Jain, Eds., Singapore: Springer Nature, 2024, pp. 253–266. https://doi.org/10.1007/978-981-97-5878-4_15.

[43] D. Castilho, D. Tedesco, C. Hernandez, B. E. Madari, and I. Ciampitti, "A Global Dataset for Assessing Nitrogen-Related Plant Traits Using Drone Imagery in Major Field Crop Species," *Scientific Data*, vol. 11, no. 1, p. 585, 2024. https://doi.org/10.1038/s41597-024-03357-2.

13 Deep Feature Extraction
Comparing VGG16, ResNet50, and InceptionV3 for Image Matching

Omaima El Bahi, Ali Omari Alaoui,
Youssef Qaraai, and Ahmad El Allaoui

13.1 INTRODUCTION

High image matching is crucial for various computer vision tasks [1] such as object recognition [2], image retrieval. and 3D reconstruction [3]. Traditional methods face challenges with real-world image complexity, leading to the adoption of deep learning, particularly convolutional neural networks (CNNs), for feature extraction [4]. CNNs learn hierarchical representations of visual data, significantly improving matching performance. This study evaluates the effectiveness of three prevalent deep learning models VGG16, ResNet50, and InceptionV3 [5] for feature extraction. Pretrained on large-scale datasets, these models have shown remarkable performance in computer vision tasks. The objective is to compare their performance in enhancing image-matching precision using a high-resolution image dataset of agricultural dams. The investigation focuses on two phases – examining feature extraction from initial and final layers of each model to identify the most informative layer, and analyzing the impact of different extraction types on matching accuracy. This contribution aims to identify the strengths and limitations of various models for image-matching tasks, guiding the development of more sophisticated strategies for real-world applications. The remainder of this chapter is organized as follows: Section 13.2 outlines related work in the field of deep learning–based image matching. Section 13.3 describes the methodology used in this study. Section 13.4 presents and discusses the experimental results. Finally, Section 13.5 concludes and discusses future research.

13.2 LITERATURE REVIEW

Deep learning has significantly advanced the field of image matching. Several studies have explored the use of CNNs for feature extraction and image matching tasks [6]. One of the pioneering works in this area is the utilization of CNNs for image classification tasks [7]. Models like VGGNet, GoogLeNet, and ResNet [5] have been pretrained on large datasets and fine-tuned for specific tasks, including image matching. Feature extraction plays a crucial role in image matching, where the goal is to identify correspondences between key points or regions in different images. Traditional handcrafted features like SIFT, SURF, and ORB [8] have been widely used for this purpose. However, deep learning–based approaches have shown promising results in learning hierarchical representations of visual data, leading to improved matching accuracy. Researchers have explored various strategies for leveraging deep CNNs for feature extraction in image matching [9]. This includes extracting features from different layers of the network, fine-tuning pretrained models on target datasets, and combining features from multiple models to enhance robustness. Recent studies have also investigated the impact of semantic and spatial features on image matching accuracy [10]. Semantic features capture high-level information about the content of images, while spatial features encode geometric relationships between key points or regions [11]. Integrating both types of features has been shown to improve matching performance, especially in challenging scenarios with significant variations in viewpoint, illumination, and occlusions. In addition to CNN-based approaches, other deep learning techniques [12] such as Siamese networks and triplet networks [13], have been proposed for learning similarity metrics directly from image pairs or triplets. These methods aim to optimize feature embedding in a way that minimizes the distance between similar images and maximizes the distance between dissimilar ones, thereby facilitating accurate image matching.

13.3 METHODOLOGY

13.3.1 FEATURE EXTRACTION

This work leverages pretrained CNNs for feature extraction in image matching. Unlike traditional methods using final classification layers, in this case the focus is on the first extraction layers within the network architecture, as shown in Figure 13.1. These layers offer a good balance between rich feature vocabulary and generalizability for matching tasks. We extract features from both first and last layers to capture a range of complexity, from basic shapes to potentially object-level information.

13.3.2 FEATURE EXTRACTION MODELS

We exploit the feature extraction qualities of three pretrained CNN architectures: VGG16 [14, 9], ResNet50 [15, 16], and InceptionV3 [17]. For VGG16, we focus on Conv1 and Conv5 layers to capture both fundamental visual features like edges and corners and more complex aspects such as object parts and textures. Res-Net50's use of residual blocks with shortcut connections facilitates efficient learning, allowing us to extract features at various stages for a hierarchy of complexity. Similarly, InceptionV3's inception modules enable capturing features at different scales and resolutions, making it adept at balancing low-level detail and broader semantic context in images.

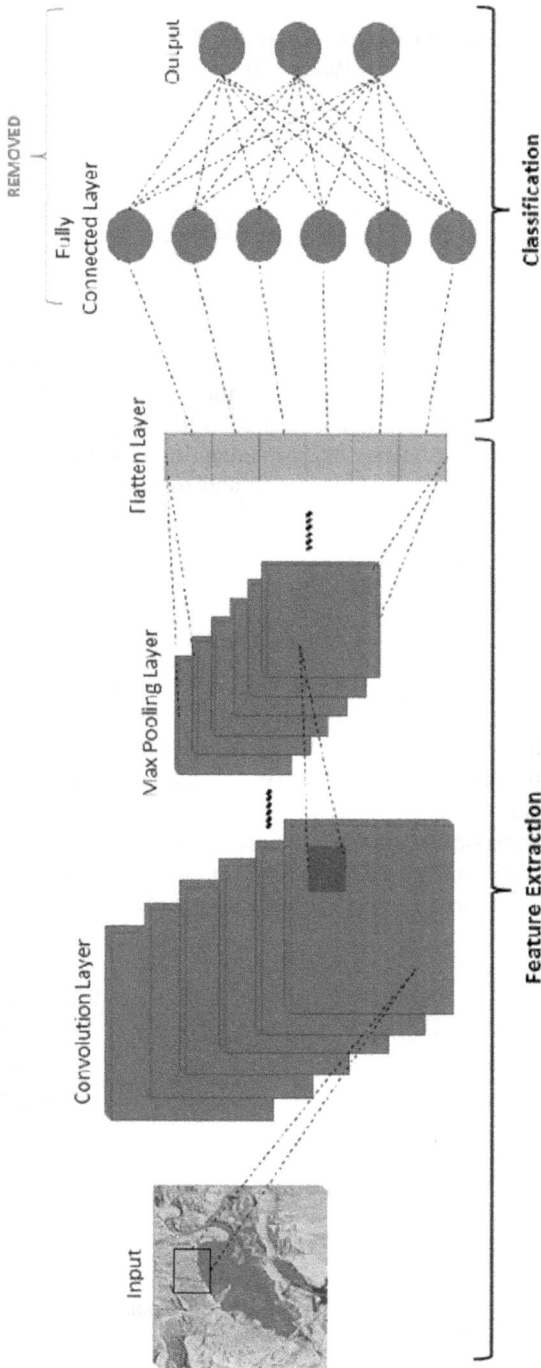

FIGURE 13.1   CNN network architecture for feature extraction.

13.3.3 Feature Matching

During the feature matching phase, we utilize the Euclidean distance metric [18] to measure the dissimilarity between features extracted from corresponding key points across multiple temporal images. The Euclidean distance, denoted as d, is calculated using the following formula [19]:

$$d = \sqrt{\sum_{i=1}^{n}\left(A_i - B_i\right)^2} \tag{13.1}$$

A_i and B_i represent the corresponding elements of feature vectors A and B, which contain features extracted from two images. To enhance robustness, features can be normalized to a unit length before distance calculation. Additionally, to ensure match reliability, a threshold is applied to filter unreliable matches and discard outliers. This threshold sets a maximum acceptable distance between key points for them to be considered valid matches. By implementing this threshold, we refine the matching process, thus boosting the accuracy and robustness of our feature-based matching approach for multi-temporal images.

13.4 RESULTS AND DISCUSSION

13.4.1 High-Resolution Dataset of Agricultural Dams

This study utilizes a dataset of high-resolution satellite images acquired from Google Earth. These images depict agricultural dams and their surrounding landscapes, providing a comprehensive view for analysis, as illustrated in Figure 13.2. To ensure consistency and comparability within the dataset, we implemented a series of pre-processing steps on the acquired images.

The dataset encompasses 63 image pairs captured between 2015 and 2023. This selection process aimed to create a diverse representation of agricultural dams within the region, facilitating a more comprehensive investigation.

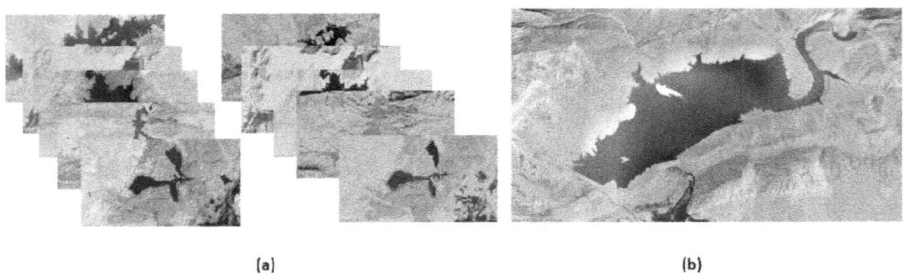

(a)   (b)

FIGURE 13.2   (a) Images of agricultural dams for correspondence analysis. (b) Satellite image representing an agricultural dam.

13.4.2 EVALUATION METRIC

We evaluated the performance and robustness of feature extraction using the VGG16, InceptionV3, and ResNet50 methods by extracting features, performing an initial match on each pair of test images and measuring accuracy [20]. Precision, a measure indicating the accuracy and reliability of our matching results [21], is calculated using the following formula:

$$Precision = TP / (TP + FP) \qquad (13.2)$$

Here, TP represents the number of true positive matches, while FP refers to the number of false-positive matches. Precision in feature-based matching refers to the ratio of correctly matched key points to all key points identified as matched. Higher precision values mean greater accuracy and reliability of matching results.

13.4.3 RESULTS AND DISCUSSION

Table 13.1 summarizes the initial results of feature-based matching accuracy tests for VGG16, InceptionV3, and ResNet50. VGG16 has the highest average accuracy (90.27%), followed by InceptionV3 (84.43%) and ResNet50 (79.07%). Interestingly, all models achieve high accuracy in certain scenarios, indicating that they are capable of accurate matching under specific conditions. While VGG16 and InceptionV3 consistently outperform ResNet50 in terms of maximum accuracy, it is worth noting that InceptionV3 may be better suited to tasks where some degree of semantic understanding is crucial, as it focuses on capturing high-level features. ResNet50, while achieving perfect accuracy in some cases, has the lowest minimum accuracy (43.59%) and the highest standard deviation (14.93%). This suggests a potential inconsistency in its matching performance between different test cases. In contrast, VGG16 and InceptionV3 have lower standard deviations, implying more stable and consistent matching accuracy.

Table 13.2 presents the results of the late feature-based matching precision test. On average, VGG16 achieves a precision of 88.24%, followed by InceptionV3 with 84.61% and ResNet50 with 82.64%. While all models demonstrate perfect precision

TABLE 13.1
Early Features-Based Matching Precision Test (Unit%)

Index	VGG16	InceptionV3	ResNet50
Average	90.27	84.43	79.07
Maximum	100.00	100.00	100.00
Minimum	60.97	59.52	43.59
Standard Deviation	09.37	10.49	14.93

TABLE 13.2
Late Features-Based Matching Precision Test (Unit%)

Index	VGG16	InceptionV3	ResNet50
Average	88.24	84.61	82.64
Maximum	100.00	100.00	100.00
Minimum	65.10	49.50	49.10
Standard Deviation	10.41	12.07	14.09

(100.00%) in certain cases, VGG16 consistently maintains the highest maximum precision across all models. However, ResNet50 exhibits the lowest minimum precision at 49.10%, indicating challenges in accurately matching key points in some instances. Moreover, ResNet50 shows the highest variability in precision with a standard deviation of 14.09%, suggesting an incoherence in performance between the different test cases. In contrast, VGG16 and InceptionV3 display lower standard deviations, implying more stable and consistent matching accuracy. Overall, VGG16 remains the preferred choice for tasks requiring accurate and reliable image matching based on late features, displaying superior performance to InceptionV3 and Res-Net50.

In feature-based early matching, VGG16 reigns supreme with an average accuracy of 90.27%. This dominance is due to its emphasis on capturing fundamental visual features such as edges and corners in early layers, which tend to be more consistent from image to image compared with higher-level features. However, the superiority of VGG16 decreases slightly in the late feature-based comparison, with an average accuracy of 88.24%. This decline can be attributed to the inherent trade-off between detail and semantic information. Later layers of convolutional neural networks capture more complex features and potentially some degree of semantic understanding. While these features can be useful for specific tasks, they do not always achieve perfect alignment between corresponding key points in image pairs, particularly if the images show significant variations in viewpoint or scene context. InceptionV3 (84.61%) and ResNet50 (82.64%) show similar trends, with a slight decrease in accuracy for matching based on late features compared with early layers.

13.5 CONCLUSION

The study presents a comprehensive comparison of the VGG16, ResNet50, and InceptionV3 models for image matching using deep feature extraction. The performance of the models has been evaluated on a high-resolution dataset of agricultural dams. Results indicate that VGG16 consistently outperforms ResNet50 and InceptionV3, particularly for early feature-based matches, where it displays superior average accuracy. Although VGG16 maintains its lead in late feature-based matching, the margin is narrowing compared to InceptionV3 and ResNet50. As a future perspective, we aim to combine early and late features extracted from deep learning models on image matching accuracy. By integrating both spatial and semantic information derived from different layers of the models, a significant improvement in feature matching precision is anticipated.

REFERENCES

[1] Jiang, X.Y., Ma, J., Xiao, G., Shao, Z., Guo, X., "A Review of Multimodal Image Matching: Methods and Applications". *Information Fusion*, 73, 22–71 (2021).

[2] Zhu, J., Wu, S., Wang, X., Yang, G., Ma, L., "Multi-Image Matching for Object Recognition". *IET Computer Vision (Print)*, 12, 350–356 (2018).

[3] Dang, W., Xiang, L., Liu, S., Yang, B., Liu, M., Yin, Z., Yin, L., Zheng, W., "A Feature Matching Method Based on the Convolutional Neural Network". *Journal of Imaging Science and Technology*, 67, 030402–030411 (2023).

[4] Yang, Z., Dan, T., Yang, Y., "Multi-Temporal Remote Sensing Image Registration Using Deep Convolutional Features". *IEEE Access*, 6, 38544–38555 (2018).

[5] Shah, S.R., Qadri, S., Bibi, H., Shah, S.M.W., Sharif, M., Marinello, F., "Comparing Inception V3, VGG 16, VGG 19, CNN, and ResNet 50: A Case Study on Early Detection of a Rice Disease". *Agronomy*, 13, 1633 (2023).

[6] Jogin, M., Madhulika, M.S., Divya, G.D., Meghana, R.K., Apoorva, S., "Feature Extraction Using Convolution Neural Networks (CNN) and Deep Learning". In *2018 3rd IEEE International Conference on Recent Trends in Electronics, Information & Communication Technology (RTEICT)* (pp. 2319–2323). IEEE (2018).

[7] Omari Alaoui, A., El Allaoui, A., El Bahi, O., Farhaoui, Y., Fethi, M.R., Farhaoui, O., "Comparative Analysis of Pre-trained CNN Models for Image Classification of Emergency Vehicles". In *The International Conference on Artificial Intelligence and Smart Environment* (pp. 256–262). Cham: Springer Nature Switzerland (2023).

[8] Bansal, M., Kumar, M., Kumar, M., "2D Object Recognition: A Comparative Analysis of SIFT, SURF and ORB Feature Descriptors". *Multimedia Tools and Applications*, 80, 18839–18857 (2021).

[9] Bahi, O.E., Alaoui, A., Qaraai, Y., Allaoui, A.E., "Deep Feature-Based Matching of High-Resolution Multitemporal Images Using VGG16 and VGG19 Algorithms". In *Lecture Notes in Networks and Systems* (pp. 516–521). Cham: Springer Nature Switzerland (2024).

[10] Ma, J., Jiang, X.Y., Fan, A., Jiang, J., Yan, J., "Image Matching from Handcrafted to Deep Features: A Survey". *International Journal of Computer Vision*, 129, 23–79 (2020).

[11] Zhang, J., Jin, X., Sun, J., Wang, J., Sangaiah, A.K., "Spatial and Semantic Convolutional Features for Robust Visual Object Tracking". *Multimedia Tools and Applications*, 79, 15095–15115 (2020).

[12] Azrour, M., et al., "A Survey of Machine and Deep Learning Applications in the Assessment of Water Quality". In *Technical and Technological Solutions Towards a Sustainable Society and Circular Economy*, J. Mabrouki and A. Mourade, Eds. (pp. 471–483), Cham: Springer Nature Switzerland (2024).

[13] Kaya, M., Bilge, H.Ş., "Deep Metric Learning: A Survey". *Symmetry*, 11, 1066 (2019).

[14] El Yanboiy, N., Khala, M., Elabbassi, I., Elhajrat, N., Eloutassi, O., El Hassouani, Y., Messaoudi, C., "Enhancing Surface Defect Detection in Solar Panels with AI-Driven VGG Models". *Data and Metadata*, 2, 81–81 (2023).

[15] Wen, L., Li, X., Gao, L., "A Transfer Convolutional Neural Network for Fault Diagnosis Based on ResNet-50". *Neural Computing and Applications*, 32, 6111–6124 (2019).

[16] Xie, S., Girshick, R., Dollár, P., Tu, Z., He, K., "Aggregated Residual Transformations for Deep Neural Networks". In *Proceedings of the IEEE Conference on Computer Vision and Pattern Recognition* (pp. 1492–1500). IEEE (2017).

[17] Liu, K., Yu, S., Liu, S., "An Improved Inceptionv3 Network for Obscured Ship Classification in Remote Sensing Images". *IEEE Journal of Selected Topics in Applied Earth Observations and Remote Sensing (Print)*, 13, 4738–4747 (2020).

[18] Malkauthekar, M.D., "Analysis of Euclidean Distance and Manhattan Distance Measure in Face Recognition". In *Third International Conference on Computational Intelligence and Information Technology (CIIT 2013)* (pp. 503–507). IET (2013).

[19] Melekhov, I., Kannala, J., Rahtu, E., "Siamese Network Features for Image Matching". In *2016 23rd International Conference on Pattern Recognition (ICPR)* (pp. 378–383). IEEE (2016).

[20] He, K., Lu, Y., Sclaroff, S., "Local Descriptors Optimized for Average Precision". In *Proceedings of the IEEE Conference on Computer Vision and Pattern Recognition* (pp. 596–605). IEEE (2018).

[21] Luo, C., Yang, W., Huang, P., Zhou, J., "Overview of Image Matching Based on ORB Algorithm". *Journal of Physics: Conference Series*, 1237, 032020 (2019).

14 The Future of Farming
AI, ML, and IoT for Improved Crop Yields

Shagufta Praveen and Mazhar Afzal

14.1 INTRODUCTION

Agriculture is one of the most important areas of our country that needs to be impro-vised through various factors. Reasons for better crop production mean better econ-omy, export revenue, rural development, environmental IoT cultural importance, and poverty reduction. In order to achieve the above factors, researchers are working on strategies that can be applied to ensure better crop production and quality crop management so that direct or indirect contributions to GDP can be increased which is crucial for economic stability [1].

Artificial intelligence (AI) is a superset of machine learning; in other words, machine learning is a subset of artificial intelligence. AI has various algorithms for different purposes, these include computer vision, pattern recognition, knowledge processing, image processing, and machine learning. Machine learning comprises mainly two learning algorithms called supervised and unsupervised [2].

Generally, learning in a machine is like asking your data to be trained and tested for processing for an accurate outcome. This is like making your machine learn how to understand data and its attributes so that related new inputs can be evaluated on the basis of last outputs. A deterministic approach can also be used for this learning which has new outputs that are evaluated on the basis of old outcomes. Other than machine learning, there are many other technologies including sensors and drones in short IoT applications are doing great in the field of agriculture and even provoked the government to come up with multiple schemes for its promotion to benefit agriculture [3, 4].

In 2020, Rashmi Priya et al. [5] proposed an idea of systematic procedures using machine learning for farming that may help create models to solve various agro-issues. In 2020, B. Ragavi proposed a seed sowing IoT-based device to reduce labor work for water generation [6]. In 2020 again, real-time soil temperature and char-acteristics were recognized by a proposed model that was IoT based [5]. In 2022, a review regarding palm tree yield prediction using machine learning was published [7]. In 2021, Naive Bayes [8], logistic regression, and a random forest algorithm were used for the prediction of crop production. In 2022, a comparative analysis was done among algorithms for crop production to measure the difference between tradi-tional and conventional farming [9]. In 2019, a decision support system was proposed to develop a system called AgroDss which helps farmers to make decisions [10].

DOI: 10.1201/9781003527664-14

In 2012, the author proposed a system using machine learning techniques that can distinguish between weeds and maize [11]. Image processing has been used to forecast crop yield for proper water management, pest control, and soil health detection [12]. Hybrid models have been created for the betterment of poor farmers [13]. An advisory system for farmers was created for the betterment of farmers [14]. The historical data like crop yields of past years can be collected from statistical yearly reports from government universities, agricultural organizations, and websites [15, 16]. Various methodological works related to soil-type clustering have been presented [17, 18]. The work presented by Ankita Patil [19] focuses on the agricultural challenges in desert areas. Sensors were deployed to collect information like temperature, soil moisture, and soil pH, so that these variables could be analyzed to maximize crop production. A methodology called linear regression with neural networks has been used to predict rainfall occurrence in relevant geographical areas [20]. For this, the data were collected from the weather station in Ahmednagar, India. and the daily maximum and minimum temperature, humidity, and rainfall for the past 10 years were studied. The linear regression and ANN helped to predict future rainfall more accurately. In Awuor et al. [21], a combination of hardware and software was presented, where sensors were used to measure soil moisture collected with the standard moisture value for a particular plant so that the amount of watering a particular plant could be monitored.

14.2 EMERGING TECHNOLOGIES AND FARMING

14.2.1 ROAD MAP FOR IMPLEMENTATION OF ML FOR CROP YIELD PREDICTION

Figure 14.1 signifies the several steps involved in crop prediction from data collection to model deployment as a web application for farmers. The role of machine

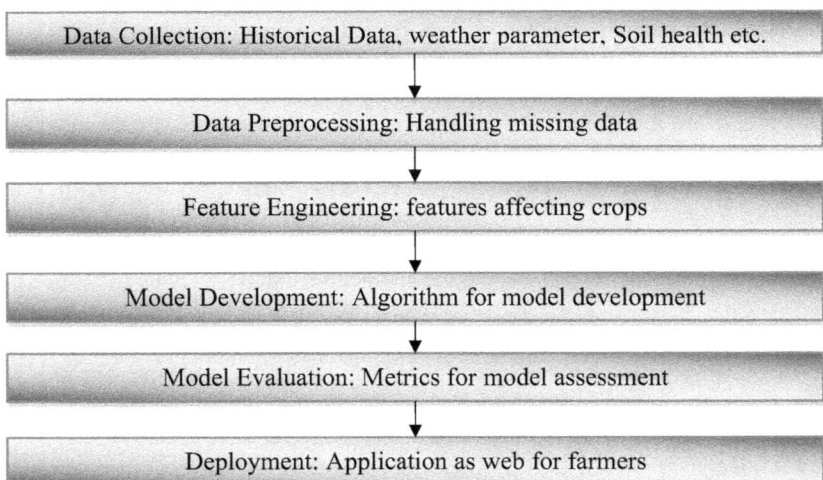

FIGURE 14.1 Road map for implementation of ML for crop yield implementation.

TABLE 14.1
ML Deployment for Crop Production

States	ML Deployment
Maharashtra	Renowned for using machine learning in pest control, water resource management, and precision farming
Karnataka	Uses machine learning in agriculture for predictive analytics and crop disease detection
Punjab	With an emphasis on yield prediction, used machine learning to optimize the production of rice and wheat
Andhra Pradesh	Uses machine learning to forecast the effects of weather on crops and to monitor soil health
Gujarat	Uses machine learning to control irrigation and forecast droughts
Uttar Pradesh	Analyzes sugarcane yield and manages pests using machine learning
Rajasthan	Focused on weather and soil data integration for sustainable farming

learning is to help farmers to get information about the crop so that they can contribute in their profession as best as they can. Almost 10 *states* are already working on ML deployments for crop production, which are depicted in Table 14.1.

14.2.2 SENSOR-BASED FARMING

Sensor-based farming, also known as precision agriculture, involves the integration of advanced technology to enhance agricultural efficiency and productivity. This approach utilizes a variety of sensors to collect critical data related to soil conditions [22], crop health, weather patterns, and other environmental factors. By gathering this information, farmers can make informed decisions regarding irrigation, fertilization, pest control, and crop rotation [23].

The process typically involves deploying sensors throughout the farming environment, which may include soil moisture sensors, temperature sensors, and satellite imaging. Once data is collected, it is shared and analyzed using sophisticated computing systems, allowing farmers to gain real-time insights into their fields. This not only optimizes resource use but also helps in predicting yield outcomes and improving overall farm management.

Additionally, sensor-based farming promotes sustainable practices by reducing waste and minimizing the environmental impact of agricultural activities. By employing data-driven strategies, farmers can enhance the quality of their crops, lower operational costs, and ultimately increase their profitability while contributing to food security (Figure 14.2). Different sensors used for farming are pH sensors, optical sensors, mechanical sensors, and so on. In Gujarat's Anand district, sensors and machine learning models are used for soil fertility analysis, with technologies like Partial Least Squares Regression being applied to predict crop yields. Punjab

| soil health monitoring | predicting best time to sow | track mineral level in soil |

| Pest Detection |

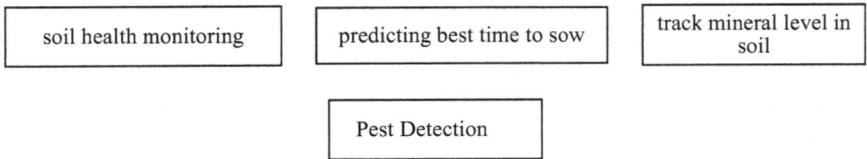

FIGURE 14.2 Application using sensor-based farming.

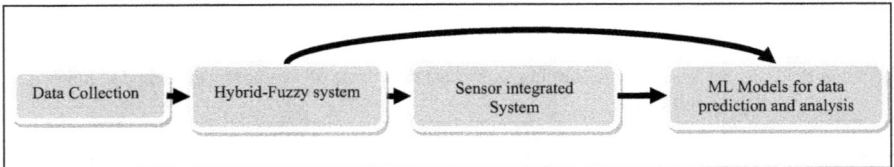

| Data Collection | Hybrid-Fuzzy system | Sensor integrated System | ML Models for data prediction and analysis |

FIGURE 14.3 Proposed model for agriculture using sensors.

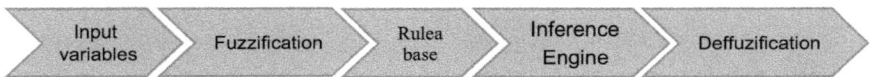

| Input variables | Fuzzification | Rulea base | Inference Engine | Deffuzification |

FIGURE 14.4 Working of fuzzy system.

has implemented sensor-based systems for monitoring diseases like yellow rust in wheat and assessing pest impacts such as the pink bollworm on cotton crops.

The proposed model (Figure 14.3) consists of collection of data from various agricultural lands. Then using the hybrid fuzzy rule system, the selection of sensors among the integrated sensors can be done and on the basis of the data, hybrid-fuzzy system will also decide to select a model to be implemented for the accurate analysis of data. As there are multiple models and multiple sensors being used for the method, the system is an automated sensor that should switch itself (Figure 14.4).

An integrated unit equipped with sensors for temperature, humidity, and pressure can effectively monitor environmental conditions by collecting real-time data. This combination of sensors allows for comprehensive climate assessments, making it ideal for applications in smart homes, agricultural practices, weather forecasting, and industrial environments. By analyzing the data from these sensors, users can optimize energy usage, enhance comfort levels, and improve overall system efficiency:

- *Inputs:* Temperature sensor accuracy = High, humidity sensor reliability = Medium, pressure sensor energy consumption = Low
- *Rule:* If temperature accuracy is high *and* humidity reliability is medium *and* pressure energy consumption is low, *then* Select temperature and pressure sensors.

- The defuzzified output may rank the temperature sensor highest, followed by the pressure sensor.

The same fuzzy rule-based approach can also be implemented for the selection of the most appropriate machine learning models by evaluating parameters such as accuracy, complexity, and training time. For instance, a fuzzy rule could state: "If accuracy is high and training time is short, then suitability is high," allowing for a nuanced assessment that accommodates the inherent uncertainties in model performance. Additional rules could incorporate other dimensions, such as "If complexity is low and accuracy is acceptable, then suitability is moderate," thereby enabling a comprehensive evaluation that considers trade-offs among various factors. This multidimensional framework would help data scientists to systematically narrow down the model choices based on a balance of efficiency and effectiveness, ultimately leading to more informed decision-making in model selection while accommodating the subjective nature of expert judgment.

14.2.3 DRONES FOR AGRICULTURE

Drones in agriculture are a modern technique of farming. Drones are a self-automated or remote-controlled systems that can be navigated using satellites or non-satellites (Figure 14.5). They have a propulsion system, cameras, and can even carry loads and travel to various areas. It is a current hero in the field of agriculture which helps in various tasks of agriculture and has provided a great contribution to crop production. The tasks are as follows:

1. Pesticide spraying
2. Spraying seeds over land
3. Using camera, the aerial area can be checked
4. Government promoting it through various advertisements and schemes
5. Crop surveillance
6. Scaring birds

FIGURE 14.5 Example of drone used in agriculture.

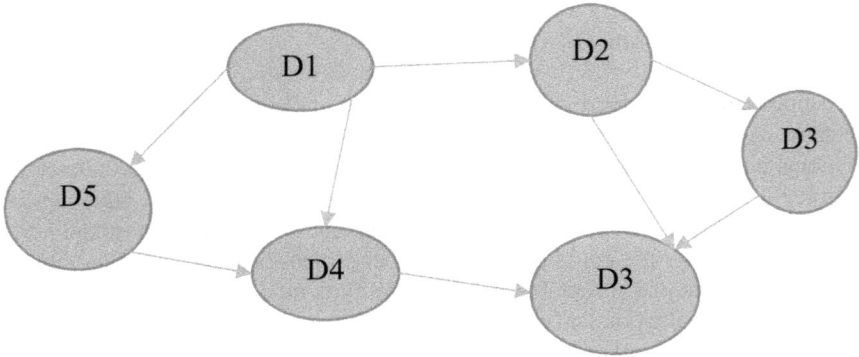

FIGURE 14.6 Drones connected as a system.

Drones consist of IoT, sensors, cameras and an advanced automation system of AI that can make decisions if it is self-automated and can do self-navigation like Tesla cars. It is one of the results of AIoT combo that is widely being used in agriculture.

- States like Punjab, Maharashtra, and Andhra Pradesh are using drones for crop health analysis and pesticide spraying.
- Organizations like the Indian Council of Agricultural Research (ICAR) and start-ups such as DeHaat and Cropin are promoting drone-based solutions [24].
- The Government of India provides subsidies and grants for adopting drone technology in agriculture under schemes like the Sub-Mission on Agricultural Mechanization (SMAM) [25].

A connected system of drones has been proposed to handle fault tolerance in agriculture (Figure 14.6). The areas which are more dependent on drones for farming can use IoT-connected systems that can inform one another so that *active–passive drones* can act according to the fulfillment or completion of the task. As drones are mechanical and there is a probability that they can be damaged during propulsion, to continue the task without delay the connected drone system can be created to provide fault tolerance. Developing a networked architecture that allows drones to cooperate and communicate in order to guarantee mission completion even in the event of individual drone failures is known as a connected system of drones for fault tolerance. To ensure operational robustness, this system uses *redundancy*, *real-time communication*, and *dynamic task distribution*.

Every drone carries a sensor that can sense an area around obstacles. In order to avoid obstacles, the connected system will go through a voting phase where through communication a drone will be appointed as a coordinator. The coordinator is the leader of the network which informed about the obstacle present and keeps the track of the drones around. If any of the drones stop working, the coordinator elects another drone to connect the team to continue with the team work.

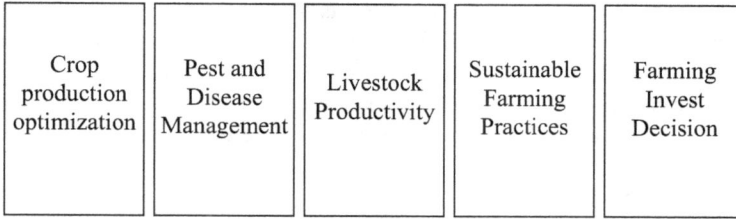

FIGURE 14.7   Decision support system.

14.2.4   Decision Support Systems for Farmers

A decision support system is an interactive, computer-based designed tool to make better decisions to improve quality of crop production using sophisticated algorithms. The decision support system generally supports agricultural management, which includes management of various components – crops, pests, livestock, and financial management.

Decision support systems are AI-based applications launched for farmers to make better decisions on various unseen situations (Figure 14.7). There are multiple offices to help, but it is not always possible to connect; so, in order to make this traditional approach more conventional, various DSSs are launched in the market to become better decision-makers in the field of agriculture or advisors for the farmers. Various algorithms for optimization and for sustainable agriculture are used for creating efficient DSSs for farmers.

The *e-SAP DSS* [26] was developed by the University of Agricultural Sciences (UAS), Bengaluru, to assist farmers in managing pest infestations and crop diseases effectively. This mobile-based system integrates real-time data collection and expert knowledge to offer precise, actionable advice to farmers.

14.3   RESULTS AND DISCUSSION

Agriculture is evolving into a more accurate, sustainable, and productive sector that incorporates contemporary technologies like machine learning (ML), drones, and decision support systems (DSSs). This chapter discusses the ML-based forecasts for weather, pest outbreaks, and soil health enabling accurate fertilizer application and ideal planting times. By detecting plant stress and identifying nutrient deficiencies early on, drones with cameras and sensors can stop crop yield loss. Labor costs are decreased via these kinds of automated systems. DSS reduces the waste of inputs like fertilizer, insecticides, and water by empowering farmers to make data-driven decisions. Precision farming maximizes irrigation and minimizes chemical runoff. Drones used for real-time monitoring cut down on waste and promote eco-friendly methods. Large regions may be covered by drones, which can provide high-resolution photographs that enable problems to be quickly identified. ML models examine this data and recommend quick fixes, like focusing on pest-control zones. High-accuracy disease and pest detection and classification is made possible by machine learning algorithms trained on plant health data, allowing for timely actions. Drones and machine learning (ML) democratize access to sophisticated farming equipment, enabling even small-scale farmers to enhance their methods. A more dependable food supply chain can result from the substantial reduction of farming guessing brought on by DSS.

14.4   CONCLUSION

This chapter wraps up the numerous facets of technology, beginning with data integration, or the flow of several data integration phases, where the most complicated step is data collecting. There are significant technological gaps for farmers as a result of the development of automatic enabled systems that require dependable Internet and possible hardware. Technical skills are necessary for improved technology implementation, yet they are frequently inaccessible in rural regions. Drones and machine learning (ML) democratize access to sophisticated farming equipment, enabling even small-scale farmers to enhance their methods. A more dependable food supply chain can result from the substantial reduction of farming guessing brought on by DSS. In order to remove financial obstacles for farmers, new technology should provide them with optimal and affordable solutions that forecast using machine learning. These new technologies promote governments for more schemes so that governments can incentivize smart farming practices through subsidies, training programs, and infrastructure development.

REFERENCES

[1] S. Dargaoui, et al., "Internet-of-Things-Enabled Smart Agriculture: Security Enhancement Approaches," in *2024 4th International Conference on Innovative Research in Applied Science, Engineering and Technology (IRASET)*, 2024, pp. 1–5. https://doi.org/10.1109/IRASET60544.2024.10548705.

[2] K. El-Moustaqim, J. Mabrouki, M. Azrour, M. Hadine, and D. Hmouni, "Enabling Smart Agriculture Through Integrating the Internet of Things in Microalgae Farming for Sustainability," in *Smart Internet of Things for Environment and Healthcare*, M. Azrour, J. Mabrouki, A. Alabdulatif, A. Guezzaz, and F. Amounas, Eds., Cham: Springer Nature Switzerland, 2024, pp. 209–222. https://doi.org/10.1007/978-3-031-70102-3_15.

[3] S. Dargaoui, et al., "IoT-Driven Smart Agriculture: Security Issues and Authentication Schemes Classification," in *Proceeding of the International Conference on Connected Objects and Artificial Intelligence (COCIA2024)*, Y. Mejdoub and A. Elamri, Eds., Cham: Springer Nature Switzerland, 2024, pp. 61–66. https://doi.org/10.1007/978-3-031-70411-6_10.

[4] J. Mabrouki, et al., "Smart System for Monitoring and Controlling of Agricultural Production by the IoT," in *IoT and Smart Devices for Sustainable Environment*, Springer, 2022, pp. 103–115.

[5] R. Priya and D. Ramesh, "ML Based Sustainable Precision Agriculture: A Future Generation Perspective," *Sustainable Computing: Informatics and Systems*, vol. 28, p. 100439, 2020. https://doi.org/10.1016/j.suscom.2020.100439.

[6] M. Ishak, M. S. Rahaman, and T. Mahmud, "FarmEasy: An Intelligent Platform to Empower Crops Prediction and Crops Marketing," in *2021 13th International*

Conference on Information & Communication Technology and System (ICTS), 2021, pp. 224–229. https://doi.org/10.1109/ICTS52701.2021.9608436.

[7] B. Ragavi, L. Pavithra, P. Sandhiyadevi, G. K. Mohanapriya, and S. Harikirubha, "Smart Agriculture with AI Sensor by Using Agrobot," in *2020 Fourth International Conference on Computing Methodologies and Communication (ICCMC)*, 2020, pp. 1–4. https://doi.org/10.1109/ICCMC48092.2020.ICCMC-00078.

[8] S. Saif, P. Roy, C. Chowdhury, S. Biswas, and U. Maulik, "Chapter 11 – Smart e-Agriculture Monitoring Systems," in *AI, Edge and IoT-based Smart Agriculture*, A. Abraham, S. Dash, J. J. P. C. Rodrigues, B. Acharya, and S. K. Pani, Eds., in Intelligent Data-Centric Systems, Academic Press, 2022, pp. 183–203. https://doi.org/10.1016/B978-0-12-823694-9.00002-5.

[9] K. Patel and H. B. Patel, "A Comparative Analysis of Supervised Machine Learning Algorithm for Agriculture Crop Prediction," in *2021 Fourth International Conference on Electrical, Computer and Communication Technologies (ICECCT)*, 2021, pp. 1–5. https://doi.org/10.1109/ICECCT52121.2021.9616731.

[10] R. Rupnik, M. Kukar, P. Vračar, D. Košir, D. Pevec, and Z. Bosnić, "AgroDSS: A Decision Support System for Agriculture and Farming," *Computers and Electronics in Agriculture*, vol. 161, pp. 260–271, 2019. https://doi.org/10.1016/j.compag.2018.04.001.

[11] J. M. Guerrero, G. Pajares, M. Montalvo, J. Romeo, and M. Guijarro, "Support Vector Machines for Crop/Weeds Identification in Maize Fields," *Expert Systems with Applications*, vol. 39, no. 12, pp. 11149–11155, 2012. https://doi.org/10.1016/j.eswa.2012.03.040.

[12] E. Khosla, D. Ramesh, R. P. Sharma, and S. Nyakotey, "RNNs-RT: Flood Based Prediction of Human and Animal Deaths in Bihar Using Recurrent Neural Networks and Regression Techniques," *Procedia Computer Science*, vol. 132, pp. 486–497, 2018. https://doi.org/10.1016/j.procs.2018.05.001.

[13] R. Priya, D. Ramesh, and E. Khosla, "Crop Prediction on the Region Belts of India: A Naïve Bayes MapReduce Precision Agricultural Model," in *2018 International Conference on Advances in Computing, Communications and Informatics (ICACCI)*, 2018, pp. 99–104. https://doi.org/10.1109/ICACCI.2018.8554948.

[14] A. Morshed, R. Dutta, and J. Aryal, "Recommending Environmental Knowledge as Linked Open Data Cloud Using Semantic Machine Learning," in *2013 IEEE 29th International Conference on Data Engineering Workshops (ICDEW)*, 2013, pp. 27–28. https://doi.org/10.1109/ICDEW.2013.6547421.

[15] "Open Government Data (OGD) Platform India." Accessed: Nov. 30, 2024 [Online]. Available: https://data.gov.in.

[16] "Home | Food and Agriculture Organization of the United Nations." Accessed: Nov. 30, 2024 [Online]. Available: www.fao.org/home/en.

[17] E. Hot and V. Popović-Bugarin, "Soil Data Clustering by Using K-Means and Fuzzy K-Means Algorithm," in *2015 23rd Telecommunications Forum Telfor (TELFOR)*, 2015, pp. 890–893. https://doi.org/10.1109/TELFOR.2015.7377608.

[18] P. Han, J. Wang, Z. Ma, A. Lu, M. Gao, and L. Pan, "Application of Fuzzy Clustering Analysis in Classification of Soil in Qinghai and Heilongjiang of China," in *Computer and Computing Technologies in Agriculture IV*, D. Li, Y. Liu, and Y. Chen, Eds., Berlin, Heidelberg: Springer, 2011, pp. 282–289. https://doi.org/10.1007/978-3-642-18333-1_33.

[19] A. Patil, M. Beldar, A. Naik, and S. Deshpande, "Smart Farming Using Arduino and Data Mining," in *2016 3rd International Conference on Computing for Sustainable Global Development (INDIACom)*, 2016, pp. 1913–1917. Accessed: Nov. 30, 2024 [Online]. Available: https://ieeexplore.ieee.org/abstract/document/7724599.

[20] M. R. Bendre, R. C. Thool, and V. R. Thool, "Big Data in Precision Agriculture: Weather Forecasting for Future Farming," in *2015 1st International Conference on Next Generation Computing Technologies (NGCT)*, 2015, pp. 744–750. https://doi.org/10.1109/NGCT.2015.7375220.

[21] F. Awuor, K. Kimeli, K. Rabah, and D. Rambim, "ICT Solution Architecture for Agriculture," in *2013 IST-Africa Conference & Exhibition*, 2013, pp. 1–7. Accessed: Nov. 30, 2024 [Online]. Available: https://ieeexplore.ieee.org/abstract/document/6701752.

[22] S. Palarimath, P. Maran, T. K, C. Balakumar, T. Sujatha, and W. B. N. R, "Exploring Sensor-Based Smart Farming Technologies in the Internet of Things (IoT)," in *2024 International Conference on Computing and Data Science (ICCDS)*, 2024, pp. 1–6. https://doi.org/10.1109/ICCDS60734.2024.10560398.

[23] P. Prokopowicz and M. Szczuka, "Hybrid Connection Between Fuzzy Rough Sets and Ordered Fuzzy Numbers," in *Fuzzy Techniques: Theory and Applications*, R. B. Kearfott, I. Batyrshin, M. Reformat, M. Ceberio, and V. Kreinovich, Eds., Cham: Springer International Publishing, 2019, pp. 505–517. https://doi.org/10.1007/978-3-030-21920-8_45.

[24] "National Conference on Building Sustainable Agricultural Startups Concluded | ICAR." Accessed: Nov. 30, 2024 [Online]. Available: https://icar.org.in/national-conference-building-sustainable-agricultural-startups-concluded.

[25] "Agriculture Drone Subsidy Scheme: Government Kisan Subsidy, License, and How to Apply Online." Accessed: Nov. 30, 2024 [Online]. Available: www.agrifarming.in/agriculture-drone-subsidy-scheme-government-kisan-subsidy-license-and-how-to-apply-online.

[26] "SAP Add-on Dynamic Safety Stock Consideration (DSS)," *SAP*. Accessed: Nov. 30, 2024 [Online]. Available: www.sap.com/documents/2017/10/a0d471b3-d97c-0010-82c7-eda71af511fa.html.

Index

Note: Page numbers in *italics* indicate a figure and page numbers in **bold** indicate a table on the corresponding page.

5G networks, 180, 198

A

accuracy, 15, 33
  dynamic texture classification, 68, 69, **69**
acoustic sensors, 77, *see also* sensors
active–passive drones, 216, *see also* drones
Adaptive Evolutionary Artificial Bee Colony-
  Back Propagation Neural Network
  (AEABC-BPNN) model, 141
Agrian Mobile app, 85
agricultural dams
  correspondence analysis, *206*
  high-resolution dataset of, 206
agricultural management platforms, 101–103, **103**
Agriculture 4.0, 183
agriculture, modern, 105, *see also* farming;
  Moroccan agriculture
  challenges, 179, 182–183
  climate change, 182
  economic pressures, 183
  environmental impact, 182
  labor shortages, 182–183
  pest and disease management, 183
  resource scarcity, 182
agriculture, conventional, 115, 182
Agrigo, 105
AgroDss, 211
Angel invariant Gabor algorithm, 78
Anticimex SMART, 86
aquaculture, 164, **166**
arable farming, 51, 182
Arduino, 51, 53–54, *54*
Area under the Receiver Operating Characteristic
  Curve (AUC–ROC), 15–16
artificial intelligence (AI), 2, 9, 10, 100, 211,
  see also explainable artificial intelligence;
  Internet of Things technology
  applications, 151
  challenges, 91–93
  data preprocessing module, 12–14
  dataset collection, **11**, 11–12
  feature distribution, 12, *12*
  feature engineering, 14
  model selection, 22–23
  in Moroccan agriculture, 99–106
  narrow and general, 76
  soil health, 114–116

for sustainable agriculture, 166–168, *167*
water quality prediction, *see* water quality
  prediction
workflow for training, evaluation, and
  selection, *11*, 14–16
Artificial Intelligence of Things (AIoT), 117
artificial neural network (ANN), 4, 33–34, 141,
  142, 195
Atmel microcontroller units (MCUs), 53
attention-based neural network (ABNN), **20**, 21
automated
  data-gathering systems, 2
  machinery and robotics, 170
  pest identification, 72, 77–79
  weather stations, 188
autonomous tractors and machinery, 196

B

Bacillus thuringiensis (Bt), 75
Bagging classifier, **20**, 21
big data analytics, 158–160
BigQuery, 87
bimodal distributions, 12–13
biological pest control, *see also* pest/disease control
  benefits, 75
  sustainable approach, 75
Bioscience newspaper, 49
blockchain with IoT (BIoT), 153, 183, 198
  applications, 157, **158**
  in different sectors, **158**
Blynk smartphone app, 51

C

cellular networks, 113
chemical-free pest control, 76, *see also* pest/
  disease control
chemical repellents, 128
climate change, 1
  agriculture and, 182
  effects on sustainable agriculture, 173
climate resilience, 172–173
Climate-Smart Agriculture (CSA) systems, 117
cloud-based platforms, 190
  data storage and processing, 190
  real-time monitoring dashboards, 190
cloud computing, 159, 160
Co-kriging techniques, 191, 195

For Product Safety Concerns and Information please contact our EU
representative GPSR@taylorandfrancis.com
Taylor & Francis Verlag GmbH, Kaufingerstraße 24, 80331 München, Germany

www.ingramcontent.com/pod-product-compliance
Lightning Source LLC
Chambersburg PA
CBHW060405220326
41598CB00023B/3020

* 9 7 8 1 0 3 2 8 6 4 6 6 2 *